Trends in the Early Careers of Life Scientists

Committee on Dimensions, Causes, and
Implications of Recent Trends in the
Careers of Life Scientists

WITHDRAWN

Board on Biology
Commission on Life Sciences
Office of Scientific and Engineering Personnel
National Research Council

National Academy Press
Washington, DC 1998

NATIONAL ACADEMY PRESS   2101 Constitution Avenue, N.W.   Washington, D.C. 20418

NOTICE: The project that is the subject of this report was approved by the Governing Board of the National Research Council, whose members are drawn from the councils of the National Academy of Sciences, the National Academy of Engineering, and the Institute of Medicine. The members of the committee responsible for the report were chosen for their special competences and with regard for appropriate balance.

This report has been reviewed by a group other than the authors according to procedures approved by a Report Review Committee consisting of members of the National Academy of Sciences, the National Academy of Engineering, and the Institute of Medicine.

This project was supported by Contract No. N01-OD-4-2139 between the National Academy of Sciences and the National Institutes of Health, Grant No. BIR-9512867 from the National Science Foundation, Grant No. APP 0589 from the Burroughs Wellcome Fund, and by.the Academy-Industry Program of the National Academy of Sciences, the National Academy of Engineering, and the Institute of Medicine. Any opinions, findings, conclusions, or recommendations expressed in this publication are those of the author(s) and do not necessarily reflect the view of the organizations or agencies that provided support for this project.

Library of Congress Catalog Card Number 98-87338
International Standard Book Number 0-309-06180-6

Additional copies of this report are available from:

National Academy Press
2101 Constitution Avenue, N.W.
Box 285
Washington, DC 20055
800-624-6242
202-334-3313 (in the Washington metropolitan area)
http://www.nap.edu

Copyright 1998 by the National Academy of Sciences. All rights reserved

Printed in the United States of America

Committee on Dimensions, Causes, and Implications of Recent Trends in the Careers of Life Scientists

Shirley Tilghman (*Chair*), Princeton University, Princeton, New Jersey
Helen S. Astin, University of California, Los Angeles, California
William Brinkley, Baylor College of Medicine, Houston, Texas
Mary Dell Chilton, Ciba-Geigy Biotechnology, Research Triangle Park, North Carolina
Michael P. Cummings, Marine Biological Laboratory, Woods Hole, Massachusetts
Ronald G. Ehrenberg, Cornell University, Ithaca, New York
Mary Frank Fox, Georgia Institute of Technology, Atlanta, Georgia
Kevin Glenn, Searle, St. Louis, Missouri
Pamela J. Green, Michigan State University, East Lansing, Michigan
Sherrie Hans, The Pew Charitable Trusts, Philadelphia, Pennsylvania
Arthur Kelman, North Carolina State University, Raleigh, North Carolina
Jules LaPidus[*], Council of Graduate Schools, Washington, DC
Bruce Levin, Emory University, Atlanta, Georgia
J. Richard McIntosh, University of Colorado, Boulder, Colorado
Henry Riecken, University of Pennsylvania (*emeritus*)
Paula E. Stephan, Georgia State University, Atlanta, Georgia

[*] until March 1997

Commission on Life Sciences Liaison
Ursula W. Goodenough, Washington University, St. Louis, Missouri

Adviser
Douglas E. Kelly, Association of American Medical Colleges, Washington, DC

Staff
Porter E. Coggeshall
Karen Greif
Charlotte V. Kuh
Alvin G. Lazen, Staff Project Director
Molla Teclemariam
James A. Voytuk
Norman Grossblatt, Editor
Kit W. Lee, Senior Project Assistant

Committee on Dimensions, Causes, and Implications of Recent Trends in the Careers of Life Scientists
Liaison Group

National Science Foundation, Arlington, Virginia
    Jim Edwards
    Joanne Hazlett
    Carlos Kruytbosch

National Institutes of Health, Bethesda, Maryland
    Jeffrey Evans
    John Norvell
    Walter Schaffer

Vanderbilt Institute for Public Policy Studies, Nashville, Tennessee
    Georgine Pion

Association of American Medical Colleges, Washington, DC
    Jennifer Sutton

Federation of American Societies for Experimental Biology, Bethesda, Maryland
    Howard Garrison

Council of Graduate Schools, Washington, DC
    Peter Syverson

The American Society for Cell Biology, Bethesda, Maryland
    Elizabeth Marincola

Commission on Life Sciences

Thomas D. Pollard (*Chairman*), The Salk Institute for Biological Studies, La Jolla, California
Frederick R. Anderson, Cadwalader, Wickersham & Taft, Washington, DC
John C. Bailar III, University of Chicago, Chicago, Illinois
Paul Berg, Stanford University, Stanford, California
Joanna Burger, Rutgers University, Piscataway, New Jersey
Sharon L. Dunwoody, University of Wisconsin, Madison, Wisconsin
John L. Emmerson, Fishers, Indiana (retired)
Neal L. First, University of Wisconsin, Madison, Wisconsin
Ursula W. Goodenough, Washington University, St. Louis, Missouri
Henry W. Heikkinen, University of Northern Colorado, Greeley, Colorado
Hans J. Kende, Michigan State University, East Lansing, Michigan
Cynthia J. Kenyon, University of California, San Francisco, California
David M. Livingston, Dana-Farber Cancer Institute, Boston, Massachusetts
Thomas E. Lovejoy, Smithsonian Institution, Washington, DC
Donald R. Mattison, University of Pittsburgh, Pittsburgh, Pennsylvania
Joseph E. Murray, Wellesley Hills, Massachusetts
Edward E. Penhoet, Chiron Corporation, Emeryville, California
Malcolm C. Pike, Norris/USC Comprehensive Cancer Center, Los Angeles, California
Jonathan M. Samet, Johns Hopkins University, Baltimore, Maryland
Charles F. Stevens, The Salk Institute for Biological Studies, La Jolla, California
John L. VandeBerg, Southwest Foundation for Biomedical Research, San Antonio, Texas

Staff
Paul Gilman, Executive Director
Alvin G. Lazen, Associate Executive Director

Board on Biology

Michael T. Clegg (*Chairman*), University of California, Riverside, California
David Eisenberg, University of California, Los Angeles, California
Gerald D. Fischbach, Harvard Medical School, Boston, Massachusetts
David J. Galas, Darwin Molecular Corporation, Bothell, Washington
David V. Goeddel, Tularik, Inc., South San Francisco, California
Arturo Gomez-Pompa, University of California, Riverside, California
Corey S. Goodman, University of California, Berkeley, California
Margaret G. Kidwell, University of Arizona, Tucson, Arizona
Bruce R. Levin, Emory University, Atlanta, Georgia
Olga F. Linares, Smithsonian Tropical Research Institute, Miami, Florida
Elliott M. Meyerowitz, California Institute of Technology, Pasadena, California
Robert T. Paine, University of Washington, Seattle, Washington
Ronald R. Sederoff, North Carolina State University, Raleigh, North Carolina
Daniel Simberloff, University of Tennessee, Knoxville, Tennessee
Robert R. Sokal, State University of New York, Stony Brook, New York
Shirley M. Tilghman, Princeton University, Princeton, New Jersey
Raymond L. White, University of Utah, Salt Lake City, Utah

Ex Officio
Thomas Pollard, The Salk Institute for Biological Studies, La Jolla, California

Staff
Paul Gilman, Acting Director
Tania Williams, Program Officer
Amy Noel O'Hara, Project Assistant

Office of Scientific and Engineering Personnel Advisory Committee

M. R. C. Greenwood (*Chair*), University of California, Santa Cruz, California
David Breneman, University of Virginia, Charlottesville, Virginia
Nancy Cantor, University of Michigan, Ann Arbor, Michigan
Carlos Gutierrez, California State University, Los Angeles, California
Stephen J. Lukasik, Los Angeles, California
Barry Munitz, California State University, Long Beach, California
Janet Norwood, The Urban Institute, Washington, DC
John D. Wiley, University of Wisconsin, Madison, Wisconsin
Tadataka Yamada, SmithKline Beecham Corporation, Philadelphia, Pennsylvania
A. Thomas Young, North Potomac, Maryland

Ex Officio
William H. Miller, University of California, Berkeley, California

Staff
Charlotte V. Kuh, Executive Director
Catherine Jackson, Administrative Assistant

The National Academy of Sciences is a private, nonprofit, self-perpetuating society of distinguished scholars engaged in scientific and engineering research, dedicated to the furtherance of science and technology and to their use for the general welfare. Upon the authority of the charter granted to it by the Congress in 1863, the Academy has a mandate that requires it to advise the federal government on scientific and technical matters. Dr. Bruce Alberts is president of the National Academy of Sciences.

The National Academy of Engineering was established in 1964, under the charter of the National Academy of Sciences, as a parallel organization of outstanding engineers. It is autonomous in its administration and in the selection of its members, sharing with the National Academy of Sciences the responsibility for advising the federal government. The National Academy of Engineering also sponsors engineering programs aimed at meeting national needs, encourages education and research, and recognizes the superior achievements of engineers. Dr. William A. Wulf is president of the National Academy of Engineering.

The Institute of Medicine was established in 1970 by the National Academy of Sciences to secure the services of eminent members of appropriate professions in the examination of policy matters pertaining to the health of the public. The Institute acts under the responsibility given to the National Academy of Sciences by its congressional charter to be an adviser to the federal government and upon its own initiative, to identify issues of medical care, research, and education. Dr. Kenneth Shine is president of the Institute of Medicine.

The National Research Council (NRC) was organized by the National Academy of Sciences in 1916 to associate the broad community of science and technology with the Academy's purposes of furthering knowledge and of advising the federal government. Functioning in accordance with general policies determined by the Academy, the Council has become the principal operating agency of both the National Academy of Sciences and the National Academy of Engineering in providing service to the government, the public, and the scientific and engineering communities. The Council is administered jointly by both academies and the Institute of Medicine. Dr. Bruce Alberts and Dr. William A. Wulf are chairman and vice chairman, respectively, of the National Research Council.

This report has been reviewed by individuals chosen for their diverse perspectives and technical expertise, in accordance with procedures approved by the National Research Council's (NRC) Report Review Committee. The purpose of this independent review is to provide candid and critical comments that will assist the authors and the Research Council in making their published report as sound as possible and to ensure that the report meets institutional standards for objectivity, evidence, and responsiveness to the study charge. The content of the review comments and draft manuscript remain confidential to protect the integrity of the deliberative process. We wish to thank the following individuals for their participa-tion in the review of this report:

Michael Clegg, University of California, Riverside, California
Marye Ann Fox, The University of Texas, Austin, Texas
Donald Fredrickson, former director, National Institutes of Health, Bethesda, Maryland
Lyle V. Jones, University of North Carolina, Chapel Hill, North Carolina
Thomas J. Kennedy Jr., formerly with the National Institutes of Health, Bethesda, Maryland, and the
  Association of American Medical Colleges, Washington, DC
William Lennarz, State University of New York, Stony Brook, New York
Jeremy Nathans, Howard Hughes Medical Institute, Baltimore, Maryland
John Perkins, University of Texas Southwest Medical Center, Dallas, Texas
Ann Peterson, W. K. Kellogg Foundation, East Battle Creek, Michigan
Thomas Pollard, The Salk Institute for Biological Studies, La Jolla, California
Ann Preston, State University of New York, Stony Brook, New York
Paul Risser, Oregon State University, Corvallis, Oregon
Lee Sechrest, University of Arizona, Tucson, Arizona
Allan Spradling, Howard Hughes Medical Institute, Baltimore, Maryland
Michael Teitelbaum, Alfred P. Sloan Foundation, New York, New York
Raymond White, University of Utah, Salt Lake City, Utah
John Wiley, University of Wisconsin, Madison, Wisconsin
William Zumeta, University of Washington, Seattle, Washington

Although the individuals listed above have provided many constructive comments and suggestions, responsibility for the final content of this report rests solely with the authoring committee and the National Research Council.

Several of the reviewers listed above have published papers on PhD workforce issues; see, for example, Dr. Kennedy's 1994 paper, "Graduate Education in the Biomedical Sciences: Critical Observations on Training for Research Careers," in *Academic Medicine* (69:10).

PREFACE

The National Research Council has regularly reported on issues of the scientific and engineering workforce, including questions related to the education, training, and deployment of scientific personnel. It actively maintains files on doctoral awardees and periodically surveys their employment in science. The Council's interest in this arena is based on the importance of scientific research to the nation's welfare, and that is also the reason for interest in support of the education and training of life scientists.

That support has chiefly come from three federal agencies: the National Institutes of Health (NIH), the National Science Foundation (NSF), and the US Department of Agriculture; numerous private foundations and public and private universities have also contributed. The US Congress has manifested interest in questions of supply of and demand for trained scientists in biomedical and behavioral science by establishing the National Research Service Award program at NIH, which provides funding explicitly for training scientists, and by requesting a periodic report from the National Academy of Sciences on national needs for biomedical and behavioral research personnel. Other agencies support life-science education and research through separate programs. Thus, this report, by the Committee on Dimensions, Causes, and Implications of Recent Trends in the Careers of Life Scientists, in the Board on Biology of the Research Council's Commission on Life Sciences, deals with issues that are pertinent to the agendas of a very wide array of agencies and institutions.

The committee was charged to examine trends in research careers of life scientists in training, at the conclusion of training, and in the years immediately after training and to examine the implication of these trends for the persons involved and for the health of the life-science enterprise. The committee's goal was to frame recommendations that would be beneficial both to the young aspirants to scientific careers and to the enterprise they had committed to. The committee recognized that it was dealing with interdependencies among educators, trainees, investigators, funders, and entrepreneurs that truly constituted a sociotechnical system of great complexity. The importance of established stakes in the status quo quickly became apparent, and the committee recognized that there was no single locus of power to make changes in the system that has produced undesirable outcomes for some young scientists. If change is to occur, it will be through the uncoordinated action of many persons at many institutions who try to consider what is best for their students and their profession and then take appropriate action.

Those insights tempered any ambition that the committee might initially have had to "reform" the system overnight by taking bold measures. The risk of doing more damage than good is great, given the complexity of the educational system, the size of the enterprise, and its importance for the nation's long-term interest. Accordingly, the committee's principal recommendations are measured rather than dramatic.

The committee appointed to prepare this report was intentionally composed of individuals with a broad range of backgrounds and experience. To be sure, 10 of them were life scientists, but their occupations and scientific expertise were diverse. Five of the 10 were tenured full professors at major universities, one a postdoctoral fellow, and one a graduate student at the time of their appointment. Two were employed in industry. Among the nonbiologists, bringing experience in studies of the scientific labor force and scientific careers and a distance from direct interest in life-science research were two economists, two psychologists, and a sociologist. The age range of the members was from the middle twenties to the middle seventies. Two department heads, a vice president for academic programs and planning, a dean of a graduate school, and a director of a research institute brought academic administrative viewpoints to the deliberations. (See appendix A for biographic sketches of the committee members.) In short, the interests of very nearly all the "stakeholders" in the life-science enterprise were represented on the committee. Such diverse outlooks richly widened the arena of discussion and were mutually educative. They also tended to slow any rush to judgment until a true consensus could be achieved.

To ensure that even the broad spectrum of views found among the committee members was enriched by outside views, we invited representatives of government and professional associations to testify before us. And we convened a public meeting at which 18 speakers presented their views and more than 50 other persons attended the meeting or made their views known through written comments. Appendix B contains the names of the speakers and other participants in this activity. A liaison group of government and scientific-organization data experts was asked to provide reactions to our early collections of data; we established contact with institutions performing relevant studies and used the information they provided. The members of the liaison group are listed after the committee roster.

Attached to this report is an alternative perspective on the committee's recommendation 3, regarding training grants. All members of the committee except the author of the alternative perspective endorsed recommendation 3 after extensive discussion at several committee meetings.

We have many other people to thank for assistance in accomplishing our task. Persons who in many instances gave up parts of their weekends to share their knowledge with the committee are Ruth Kirschstein, Walter Schaffer, John Norvell, and James Onken, of NIH; Mary Clutter and Joanne Hazlett, of NSF; Douglas Kelly, Jennifer Sutton, and Stanley Ammons, of the Association of American Medical Colleges (AAMC), Mary Jordan, of the American Chemical Society; and Roman Czujko, of the American Institute of Physics. Participants in and contributors to our public meeting are listed in appendix B. Walter Schaffer, of NIH, and James Edwards, of NSF, were extremely helpful in their roles as program officers on behalf of their agencies. Data were made available by and useful discussions held with John Norvell, of NIH; Lawrence Burton of NSF; Lisa Sherman and Brooke Whiting, of AAMC; Georgine Pion, of Vanderbilt University; and Thomas J. Kennedy Jr. Edward O'Neill and Renee Williard, of the University of California, San Franciso (UCSF) Center for the Health Professions, provided us with their report on Pew scholars in the biomedical sciences, and the BioMedical Association of Stanford University, and the Postdoctoral Scholars Association of UCSF shared the results of their surveys of graduate students and postdoctoral fellows.

The committee's task would have been immeasurably harder without the constant logistic, managerial, and professional support of Al Lazen, Porter Coggeshall, James Voytuk, Karen Greif, Charlotte Kuh, and Molla Teclemariam. At every stage of our work, these dedicated National Research Council staff prepared material for our enlightenment, responded to requests for more help, and took a constructive part in our meetings; they deserve no blame and much credit for our report.

Shirley Tilghman
Chair, Committee on Dimensions, Causes,
and Implications of Recent Trends in the
Careers of Life Scientists

Contents

Executive Summary 1

Chapter 1 Introduction 11

Chapter 2 Education and Research Training of Life-Science PhDs 21

Chapter 3 Early-Career Employment Profiles of Life-Science PhDs 33

Chapter 4 Opportunities, Constraints, and Future Needs 49

Chapter 5 Implications of the Findings 63

Chapter 6 Conclusions and Recommendations 79

Alternative Statement on Recommendation 3: Henry Riecken 91

Appendixes

A Biographic Information 95

B Participants in Public Meeting 97

C Sources of Data 99

D Doctoral Fields Included for Data Analysis 103

E Data Tables for Chapter 2 105

F Data Tables for Chapter 3 129

G Getting Started on the World Wide Web: Web Sites of Interest to Young Scientists 175

EXECUTIVE SUMMARY

The 50 years since the end of World War II have seen unprecedented growth in the life sciences. In 1997 US government investments in health research exceeded $14 billion, private foundations contributed more than $1.2 billion, and industry's investment in health research and development exceeded $17 billion. Government and private support of agriculture and environmental research approached $5 billion. Clearly, the life-science enterprise is large and vigorous.

The large investment in the life sciences has produced many important results. Discoveries in agricultural science have improved our understanding of soils and their chemistry and have led to the development of new strains of crop plants that are resistant to diseases and yield more food per cultivated acre. Environmental sciences and forestry have evolved new methods for managing sustainable resources that will help our expanding population to pass on more of its natural wealth to future generations. Medical science has provided fundamental understanding of the molecular basis of numerous diseases which has led to the elimination of some and the containment of many. Advances in molecular biology not only have spawned the economically important biotechnology industry but have contributed fundamental knowledge about the structure of genes and the behavior of biological macromolecules. These advances have yielded new insights into the relationships among organisms and into the continuum of structure and function that connects living and nonliving things. The long-range implications of all the rapidly evolving knowledge are hard to predict, but many additional benefits are now on the horizon.

The continued success of the life-science research enterprise depends on the uninterrupted entry into the field of well-trained, skilled, and motivated young people. For this critical flow to be guaranteed, young aspirants must see that there are exciting challenges in life-science research and they need to believe that they have a reasonable likelihood of becoming practicing independent scientists after their long years of training to prepare for their careers. Yet recent trends in employment opportunities suggest that the attractiveness to young people of careers in life-science research is declining.

In the last few years, reports from the National Research Council have detailed a changing world for young scientists. A 1994 study sought to determine whether young investigators in the biologic and biomedical sciences might be at a disadvantage compared with older, established scientists in the competition for research support. The study found no evidence of discrimination by age in National Institutes of Health (NIH) awards; but it did reveal that NIH research-grant applications from investigators below the age of 37 had plummeted between 1983 and 1993. The reasons for the decline were not immediately obvious, but concern over the seeming contraction of young research talent led to the appointment of the present committee. A 1995 study examined graduate education in all fields of science and engineering and the changing employment opportunities for PhD graduates. That committee found that more than half of new graduates with PhDs in all disciplines now find employment in nonacademic settings, and it recommended that graduate programs diversify to reflect the changing employment opportunities afforded PhD scientists.

This report extends the analyses of the previous reports by examining the changes that have occurred over the last 30 years in graduate and postgraduate training of life scientists and the nature of their employment on completion of training. It suggests reasons for the decrease in the number of young scientists applying for NIH grants and the growing "crisis in expectation" that grips young life scientists who face difficulty in achieving their career objectives.

CHARGE

This committee was charged to:
- Construct a comprehensive data profile of the career paths for recent PhD recipients in the life sciences.
- Use the profile for assessing the implications of recent career trends for individuals and for the research enterprise.
- Make recommendations, as appropriate, about options for science policy.

The charge called on the committee to consider all the life sciences and the health of the enterprise in addition to the well-being of the individuals involved.

The committee approached its first task by analyzing data contained in the large databases maintained by the National Research Council Office of Scientific and Engineering Personnel, which provides the most comprehensive and accurate record available of the educational course and employment status of scientists educated to the PhD level in the United States. These records are collected when degrees are awarded and updated biennially through surveys of a sample of doctorate holders. The committee's analysis included the 1970-1995 surveys, and thus enabled documentation of trends in important career stages.

FINDINGS

The training and career prospects of a graduate student or postdoctoral fellow in the life sciences in 1998 are very different from what they were in the 1960s or 1970s. Today's life scientist will start graduate school when slightly older and take more than 2 years longer to obtain the PhD degree. Today's life-science PhD recipient will be an average of 32 years old. Furthermore, the new PhD today is twice as likely as in earlier years to take a postdoctoral fellowship and thus join an ever-growing pool of postdoctoral fellows—now estimated to number about 20,000—who engage in research while obtaining further training and waiting to obtain permanent positions. It is not unusual for a trainee to spend 5 years—some more than 5 years—as a postdoctoral fellow. As a consequence of that long preparation, the average life scientist is likely to be 35-40 years old before obtaining his or her first permanent job. The median age of a tenured or tenure track faculty member is now about 8 years more than that of the faculty member of the 1970s.

Those facts suggest one source of the seeming contraction of "young investigator" applicants for NIH research grants. "Young" investigators have grown older, and fewer are in faculty positions before the age of 37. More of them are postdoctoral fellows, who, by most institutional regulations, may not submit applications for individual research grants.

There have been major changes in career opportunities for PhDs over the last 3 decades. Historically, the three major employment sectors for life scientists have been academe, industry, and government; academe has been the largest. The opportunity to secure an academic appointment has steadily narrowed since the 1960s. Of life scientists who received the PhD in 1963 and 1964, 61% had achieved tenured appointments at universities or 4-year colleges 10 years later. For the cohort who graduated in 1971-1972, that percentage had dropped to 54%; and for the 1985-1986 cohort, to 38%. The probability of industrial employment rose from 12% to 24% for the cohorts described above, and the probability of working in a federal or other government laboratory dropped from 14% to 11%. Overall, the fraction of PhDs with "permanent"[1] positions in the traditional employment sectors for PhD

---

[1] The committee defines the goal of graduate education and postdoctoral training in the life sciences as the preparation of young scientists for careers as independent researchers in academe, industry, government, or a private research environment. We call positions in those careers "permanent", although it is understood that no employment is guaranteed, to distinguish them from the "impermanent" positions, such as postdoctoral and research associate-positions held by persons whose career objective is to obtain permanent positions.

Executive Summary

scientists–academe, industry, and government–9-10 years after receipt of the PhD declined from 87% to 73% from 1975 to 1995. For the cohort 5-6 year after receipt of the PhD, the fraction has declined from 89% to 61% from 1975 to 1995.

During most of the time that those changes in permanent research-career outcomes were taking place, the number of life-science PhDs awarded annually by American universities was growing steadily, but slowly, from about 2,700 in 1965 to about 5,000 in the middle 1980s. Then, in 1987, the number began to rise rather steeply–to 7,696 in 1996. PhDs awarded to foreign nationals made up the majority of the increase after 1987. There has been a steady increase in the number of women receiving PhDs since 1965. Differences exist between biomedical and nonbiomedical fields; almost all the growth in numbers among life-science PhDs has been in the biomedical fields.

The 42% increase in PhD production between 1987 and 1996 was not accompanied by a parallel increase in employment opportunities, and recent graduates have increasingly found themselves in a "holding pattern" reflected in the increase in the fraction of young life scientists who after extensive postdoctoral apprenticeships still have not obtained permanent full-time positions in the life sciences. In 1995, as many as 38% of the life-science PhDs–5-6 years after receipt of their doctorates–still held postdoctoral positions or other nonfaculty jobs in universities, were employed part-time, worked outside the sciences, or were among the steady 1-2% unemployed. The comparable fraction in 1973 was only 11%. What may be most alarming about the 1995 figure is that it reflects the situations of those earning PhDs in 1989 and 1990, at the beginning of the sharp rise in the rate of PhD production.

The frustration of young scientists caught in the holding pattern is understandable. These people, most of whom are 35-40 years old, typically receive low salaries and have little job security or status within the university. Moreover, they are competing with a rapidly growing pool of highly talented young scientists–including many highly qualified foreign postdoctoral fellows–for a limited number of jobs in which they can independently use their research training. This situation–and its implications for both individual scientists and the research enterprise–is a matter of concern to the committee.

The committee viewed it as unlikely that conditions will change enough in the near future to provide employment for the large number of life-science PhDs now waiting in the holding pattern. Federal funding for life-science research is expected to grow but the growth is unlikely to compensate for the imbalance in production of PhDs as federal funding was growing substantially through the 1980s and 1990s while the employment situation for the increasing number of young life graduates worsened. We believe that the growth in funding does not ensure that trends in obtaining permanent jobs will improve. The cost of doing research at private universities has been borne traditionally by federal and private granting agencies, and it is highly unlikely that tuition, already high, can be increased to the extent that it could provide needed research support. Schools of medicine, where large numbers of PhDs are educated and work, are faced with the need to adjust to the era of "managed care" with a marked reduction in income from clinical-practice plans that previously contributed substantially to the support of research and training. Finally, industry–and perhaps specifically the biotechnology sector–which employed large numbers of new life-science PhDs in the 1980s, has slowed its hiring in the 1990s.

In response to the increasing difficulty of finding employment in traditional sectors, trainees and their mentors have looked to alternative careers, such as law, science writing, science policy, and secondary-school teaching. Our analysis suggests that opportunities in these fields might not be as numerous or as attractive as advocates of alternative careers imply.

IMPLICATIONS

Whether the career trends described above are a source of concern depends on the viewpoint of the stakeholder observing them. To the graduate student and postdoctoral trainee who have invested many years of preparation with the expectation of having a research career, the situation is discouraging indeed. To the established investigator and the overseers of life-science research, the availability of large numbers of bright young scientists willing to work very hard for relatively little financial compensation is an asset that contributes to a remarkably successful enterprise. Since World War II, the structure of life-science research has been built around these young scientists, who are the primary producers of research. The public, whose taxes support the enterprise, has benefited from the abundance of young trainees.

The imbalance between the number of life-science PhDs being produced and the availability of positions that permit them to become independent investigators concerns the committee. The long times spent in training, the delay in achieving independence, and especially the difficulty in finding positions where young scientists can independently use their training have led to a "crisis in expectation". The feelings of disappointment, frustration, and even despair are palpable in the laboratories of academic centers. Many graduate students entered life-science training with the expectation that they would become like their mentors: they would be able to establish laboratories in which they would pursue research based on their own scientific ideas. The reality that now faces many of them seems very different. The future health of the life sciences depends on our continuing to attract the most talented students. That will require that students be realistically informed at the beginning of their training of their chances of achieving their career goals and that faculty recognize that current employment opportunities are different. The challenges for the life-science community are to acknowledge that it is the structure of the profession that has led to declining prospects for its young and to develop accommodations to maximize the quantity and quality of the scientific productivity of the future.

CONCLUSIONS AND RECOMMENDATIONS

The committee's analysis of the patterns of employment of recent recipients of life-science PhDs suggests that the current level of PhD production now exceeds the current availability of jobs in academe, government, and industry where they can independently use their training. While only a small minority of recent PhDs have left the field entirely, a large fraction of the "excess" supply is currently found in two kinds of appointments, "postdoctoral" and "other academic," where they may be called "fellows", "research assistants", "adjunct instructors" or some other title that conveys a clear message of impermanence in academe–outside the tenure track of regular faculty.

The professional structure of the life sciences research enterprise, in which the important work of conducting experiments rests almost entirely on the shoulders of graduate students and postdoctoral fellows, was based on the premise that there would be continuous expansion of available independent research positions in order to provide employment commensurate with their training for the ever-growing number of trainees. By the 1980s, however, there were signs of trouble ahead as the postdoctoral pool began to swell in size. The dramatic jump in number of graduates from PhD programs that began in 1987, driven by the influx of foreign-born PhD candidates together with the increase in foreign-trained PhDs who have sought postdoctoral training in the US, has greatly exacerbated what was already the growing imbalance between the rate of training versus the rate of growth in research-career opportunities.

Although the current abundance of PhDs is an advantage to established investigators, those responsible for graduate education in the life sciences should realize that further growth in the rate of PhD training could adversely affect the future of the research enterprise. Intense competition for jobs has created a "crisis of expectation" among young scientists; further increase in the competition could discourage the best from entering the field.

Recommendation 1: Restraint of the Rate of Growth of the Number of Graduate Students in the Life Sciences

The committee recommends that the life-science community constrain the rate of growth in the number of graduate students, that is, that there be no further expansion in the size of existing graduate-education programs in the life sciences and no development of new programs, except under rare and special circumstances, such as a program to serve an emerging field or to encourage the education of members of underrepresented minority groups.

The current rate of increase in awards of life science PhDs–5.1% from 1995 to 1996–if allowed to continue, would result in a doubling of the number of such PhDs in just 14 years. Our analysis suggests that would be deleterious to individuals and the research enterprise. The committee recognizes that the number of PhDs awarded each year might already be too high. Although a return to pre-1988 levels of training might be beneficial, we believe that a concentrated effort to reduce the size of graduate-student populations rapidly would be disruptive to the highly successful research enterprise. While our data show a current abundance, some unanticipated discovery in the life sciences or unexpected change in funding trends might lead to an increase in demand for life scientists. The committee believes that the current system is well prepared to meet such a need.

We caution that it will be necessary to distinguish among fields when making decisions about optimal numbers of graduate students. As shown in chapter 2, almost all the increase in life-science PhD production has been in biomedical fields. Actions taken in one field of the life sciences might be unnecessary in others. It is worth noting, however, that the data shown in figure 3.10 suggest that biomedical and nonbiomedical life-science fields are experiencing similar changes in employment trends, for example, smaller fractions of PhDs finding permanent employment in academe.

The committee examined several approaches to stabilizing the total number of PhDs produced by life-science departments beyond the first and obvious approach of individual action on the part of graduate programs to constrain growth in the number of graduate students enrolled. Some might argue that this solution is expecting unreasonably altruistic behavior on the part of established investigators and training-program directors and that graduate programs will continue to accept large numbers of students simply to meet their faculties' need for instructors and laboratory workers. The committee urges life-science faculties to seek alternatives to these workforce needs by increasing the number of permanent laboratory workers. As the increases over the last decade have been fueled almost entirely by the increased availability of federal and institutional support for research assistants, consideration might be given to restricting the numbers of graduate students supported through the research-grant mechanism.

The committee believes the most prudent way to reasonably reduce the rate of increase in the number of PhDs awarded annually and perhaps to achieve a gradual reduction in the numbers being trained is to help students to make informed decisions about their career choices.

To be effective, such decisions must be based on relevant and up-to-date information about both the quality of the training available in particular graduate programs and in the job opportunities of a given field. Equally importantly, this information must be used by individual graduate programs and mentors in determining the numbers of trainees they accept and in assessing the effectiveness of their programs. It is our expectation that such information will have the salutary effect of letting market forces control the rate of entry into the profession *before* young people have invested ten and more years in training.

Recommendation 2: Dissemination of Accurate Information on the Career Prospects of Young Life Scientists

> **The committee recommends that accurate and up-to-date information on career prospects in the life sciences and career outcome information about individual training programs be made widely available to students and faculty. Every life science department receiving federal funding for research or training should be required to provide to its prospective graduate students specific information regarding all predoctoral students enrolled in the graduate program during the preceding 10 years.**

With the most accurate information available, students will be able to make informed decisions about their careers.

Recommendation 3: Improvement of the Educational Experience of Graduate Students

There is no clear evidence that career outcomes of persons supported by training grants are superior to those of persons supported by research grants. However, the committee, which included members with direct experience with training grants, concluded that training grants are pedagogically superior to research grants and result in a superior educational climate in which students have greater autonomy. First, training grants are pedagogically superior because they provide a mechanism for stringent peer review of the training process itself, something that is not considered in the review of a research project. Second, they improve the educational climate because they minimize the potential conflicts of interest that can arise between trainers and trainees. Although the student-mentor relationship is ordinarily healthy and productive for both partners, it can be distorted by the conditions of the mentor's employment of the student and limit the ability of students to take advantage of opportunities to broaden their education. Third, training grants provide the federal government with information that it needs to evaluate the level of its investment in graduate life-science education with the aim of developing a funding framework for graduate education that contributes to the long-term stability and well-being of the research enterprise.

> **The committee encourages all federal agencies that support life-science education and research to invest in training grants and individual graduate fellowships as preferable to research grants to support PhD education. Agencies that lack such programs should look for ways to start them, and agencies that already have them should seek ways to sustain and in some instances expand them.**

> **This recommendation should not be pursued at the expense of scientific and geographic diversity. Rather, we en-**

courage the establishment of small, focused training-grant programs for universities that have groups of highly productive faculty in important specialized fields, but might not have the number of faculty needed for more traditional, broad-based training grants.

It is true that the current regulations governing NIH training grants bring universities some financial disadvantages because of restricted overhead recovery. Furthermore, NIH training grants cannot support foreigners on student visas, and so this recommendation places at disadvantage programs that depend on foreign students for research or teaching. These disadvantages are outweighed, in the committee's view, by the salutary effect that the training-grant peer-review process brings to the members of a department faculty, leading them to examine and reflect on how, as an entity, they are providing for the education and training of their graduate students.

Our endorsement of training grants and fellowships is not intended to result in the training of more PhDs. Rather we advocate a shift from support by research grants to training grants. We anticipate improvements in the quality and oversight of graduate education in the life sciences. The federal government is already heavily invested in life-science education; greater reliance on support of graduate students on training grants ensures that taxpayers are receiving the best return on their investment.

The committee is also concerned that the length of time spent in training has become too long at a median of 8 years elapsed time from first enrollment to PhD for all life sciences (though field differences exist). We believe that the time should be about 5-6 years. However, an immediate effort to shorten the time to degree would increase the number of PhDs produced. Efforts to shorten the time to degree should be undertaken when the effort to restrain growth in the number of PhDs has shown positive effects.

Recommendation 4: Enhancement of Opportunities for Independence of Postdoctoral Fellows

While the length of graduate training has been increasing, so too have the extent and duration of postdoctoral training. Prolonged tenure as a postdoctoral fellow provides a person with valuable research experience, but it carries some real costs. In most cases, fellows are not independent of their mentors so they can not pursue their own research. We recognize the many good reasons for prolonged tenure as a postdoctoral fellow but we believe that tenures longer than 5 years are not in the best interest of either the individual fellow or the scientific enterprise.

Because of its concern for optimizing the creativity of young scientists and broadening the variety of scientific problems under study in the life sciences the committee recommends that public and private funding agencies establish "career-transition" grants for senior postdoctoral fellows. The intent is to identify the highest-quality scientists while they are still postdoctoral fellows and give them financial independence to begin new scientific projects of their own design in anticipation of their obtaining fully independent positions.

The committee recommends a goal of 200 federal and private grants awarded annually, representing about 1% of the postdoctoral pool. That number of people supported would be quite small but the program might provide an important opportunity for the most

promising postdoctoral fellows and serve as both example and incentive to many more. We make this recommendation with the knowledge that it is possible that the money for a new federal grant program probably would come from existing federal funds. In our view, the benefits of increased intellectual independence and improved motivation of talented midcareer postdoctoral fellows justify such a reallocation of funds. Private funders might establish new programs or enlarge existing programs that support career-transition grants.

Recommendation 5: Alternative Paths to Careers in the Life Sciences

As traditional research positions in academe, industry, and government have become more difficult to obtain, positions in "alternative careers"—such as law, finance, journalism, teaching, and public policy have been suggested as opportunities for PhDs in the life sciences.

The idea of highly trained scientists investing their talents in nontraditional careers seems at first glance attractive. Scientists have analytical skills and a work ethic to bring to any position, and the placement of highly trained scientists in diverse jobs in the workforce would lead to an increase in general science literacy. As the committee's review of alternative opportunities concludes, however, most of the possibilities are less available or less attractive than they might at first glance appear. Many "alternative" careers are also heavily populated, and competition for good positions is stiff. Others require special preparation or certification, or offer unattractive compensation, and none makes full use of the PhD's hard won life-science research skills. The committee believes that the idea of alternative careers should not be oversold to PhD candidates.

The interest in alternative careers for PhD scientists has inevitably raised the question of whether preparation for the degree should be changed from its current narrow focus on training for the conduct of scientific research to embrace a broader variety of educational goals that would connect to alternative career paths. The committee has discussed that question extensively.

The committee recommends that the PhD degree remain a research-intensive degree, with the current primary purpose of training future independent scientists.

At the same time, the committee recognizes that not all students who begin graduate school intending to pursue a research career maintain that desire as they progress through training. Graduate programs should expand their efforts to help students learn about the diversity of career opportunities open to them, and university departments should examine possible alternatives to the research PhD.

One alternative to broadening the PhD program is to strengthen the Masters degree, which may be a more appropriate end point for students who determine early enough in their training that PhD training is not necessary for the career goals they have selected. There has been a decline in the number of master's-degree programs in the life sciences and with it a growing perception that the master's degree has become a consolation prize for those who do not complete a PhD program. This devaluation of the master's degree effectively limits the number of choices for college graduates who are interested in a career in the life sciences, although not necessarily careers in directing laboratories conducting fundamental research. For example, the college graduate who is interested in teaching in secondary school or two-year colleges, would benefit from formal and focused master's-degree programs that do not require long periods of research-intensive graduate and postdoctoral training. Masters degree programs would not only be more appropriate but also be preferable to the PhD for this type of employment and these students.

We recommend that universities identify specific areas of the biological and biomedical sciences for which Master's level training is more appropriate, more efficient and less costly than PhD training. We recommend that focused Master's Programs be established in those areas.

A vigorous master's-degree program that produces highly skilled laboratory technicians for industry, government, and academe could potentially contribute to righting the imbalance between PhD training and the labor market. When the committee recommended constraint in further growth in training in recommendation 1, it was fully aware that graduate students are needed in the labor-intensive life-science research enterprise and to teach undergraduates. One way to resolve this dilemma is to effect a modest shift toward a more permanent laboratory workforce by replacing some fraction of the existing training positions with permanent employees such as MSc-level technicians and PhD-level research associates.

The Impact of Foreign Nationals

This report has documented that the majority of the recent increase in the number of PhD trainees and postdoctoral fellows are foreign nationals, not US citizens. The number of foreign nationals reflects the international nature of modern science and the central place that the US plays in this international arena. Furthermore, foreign nationals have traditionally contributed to the excellence of US science, as suggested by the fact that of the 732 members of the National Academy of Sciences who are life scientists, 21.2% are foreign born and 12.4% obtained their PhD training abroad. Foreign nationals' important contributions to US scientific leadership is reflected in their representation as department chairs (25%) and their inclusion as "outstanding authors" in life sciences (26.4%). Foreign students and fellows are welcome participants in the research enterprise, provided they are of high quality and competitive with American applicants.

We believe it would be unwise to place arbitrary limitations on the number of visas issued for foreign students. But we do not believe that US institutions should continue to enroll unlimited numbers of foreign nationals. As decisions are made on ways to constrain further growth, the measures adopted should apply equally to all students regardless of nationality.

If, as we hope, implementation of our recommendations results in constraining further growth in PhDs awarded in the life sciences, we urge our colleagues on graduate admissions committees to resist the temptation to respond by simply increasing the number of foreign applicants admitted.

Responsibility for Effecting Change

This report has documented several dramatic changes in career trends in the life sciences over the last several decades. The rapid growth in the academic scientific establishment in the 1960s and the early 1970s set in place a training infrastructure that was built on the premise that there would be continued growth. When the inevitable slowdown in resources to support that growth occurred, it was not accompanied by a commensurate adjustment in the rate of training. The impact of the imbalance between the number of aspirants and the research opportunities is now being felt by a generation of scientists trained in the last 10 years who are finding it increasingly difficult to find permanent positions in which their hard-accumulated skills in research can be used. Unless steps are taken to put the system more in balance, the difference between students' expectations and the reality of the employment market will only widen and the workforce will become more disaffected. Such an occurrence would damage the life-science research enterprise and all the participants in it.

The training of life scientists is a highly decentralized activity. Notwithstanding the heavy dependence on federal funds, the most important decisions affecting the rate of production of life scientists are made locally by the universities and their faculties. The numbers and qualifications of students admitted to graduate study, the allocation of institutional funds for their tuition and stipends (which account for half or more of the total expenditures for graduate-student support), the requirements for the degree—all are local decisions. As a consequence, a large portion of the responsibility for implementing our recommendations falls on the shoulders of established investigators, their departments and universities, professional scientific organizations, and students themselves. Students must take the responsibility of making informed decisions about graduate study, but they must be provided accurate career information on which to base their decisions. Individual faculty members must be willing to set aside their short-term self-interest in maintaining the high level of staffing of their laboratories for the sake of the long-term stability and well-being of the scientific workforce. Directors of graduate programs must be willing to examine the future workforce needs of the scientific fields in which they train, not just the current needs of their individual departments for research and teaching assistants.

The recommendations in this report are offered as first steps to improve the overall quality of training and career prospects of future life scientists. We hope that the information in this report will be used to begin discussions within the life-science community on the best ways to prepare future scientists for exciting careers in the profession and to protect the vitality of the life-science research enterprise.

1 INTRODUCTION

A CAPSULE HISTORY OF AMERICAN RESEARCH IN THE LIFE SCIENCES

During the latter half of the 20th century, the United States has witnessed substantial growth in the size and effectiveness of its life-science research enterprise. Indeed, the very definition of life science has emerged during this century as the sum of agricultural, biochemical, cellular, developmental, ecologic, evolutionary, molecular, and medical biology. The National Institute of Health was established by the Ransdell Act in 1930 (PL 71-251), but during the 1930s life-science research in university and industry laboratories was conducted with little support from the government. The US Department of Agriculture (USDA) was the only source of federal support for such work. The National Cancer Institute (NCI) was established in 1937, but although its mandate included the funding of research and training in nonfederal laboratories, its expenditures for medical research in 1940 were only $3 million, including both intramural and extramural work. Meanwhile, private sources, such as the Rockefeller Foundation, contributed $17 million, and industry $25 million (NIH 1961). In 1944, Congress pluralized the National Institutes of Health (NIH) to include several disease-oriented institutes in addition to NCI, but at no time between 1938 and 1945 did NIH extramural expenditures exceed $250,000 (NIH 1978).

In the period before World War II, the number of life scientists trained per year was also low; in 1930, only 342 PhDs were awarded in all the life sciences. By 1940, however, change was in the air: Warren Weaver, of the Rockefeller Foundation, noted that "gradually there is coming into being a new branch of science–molecular biology–which is beginning to uncover many secrets . . . of the living cell" (Judson 1979), and the number of life-science PhDs awarded was 672. It was, however, the events during and after World War II that had the greatest effect on the climate of life-science research. The pressing problems of wartime required solutions on an unprecedented scale. Whole armies became ill with malaria, and drugs for the treatment of infection and trauma were needed in massive amounts. Rates of food production became an issue of international importance. For the first time, life scientists were mobilized on a broad front and given abundant resources with which to tackle the fundamental and practical problems of biology; and both medical and agricultural problems were solved. The successes of those efforts and of comparable work in other fields of science gave credibility to the idea that the entire United States could benefit from institutionalized support for research, as propounded in the 1945 report by Vannevar Bush, *Science, the Endless Frontier* (NSF 1960).

The postwar years saw the establishment of the National Science Foundation (NSF) and an expansion of NIH. By 1947, the government was investing $28 million per year in medical research, 9 times the investment of 7 years earlier and approaching industry's $35 million. By 1960, NSF was spending $29 million on biologic and medical sciences. From 1956 to 1961, NIH expenditures for extramural research went from $40.5 million to $272.9 million; during the same period, NIH investments for training grew from $17.3 million to $132 million, proportionally an even larger increase (NIH 1961), so funds for training kept pace with support for research. Indeed, an important consequence of Bush's blueprint for federal investment in science was the establishment of a linkage between research and research training. It was a natural consequence of the policy that federally supported research would be conducted primarily in university-based research laboratories. As the funds for research increased in the postwar years, the number of life-science PhDs granted per year grew correspondingly–from 1,660 in 1960 to 4,980 in 1971, tripling in only 10 years.

Those patterns of government investment had profound effects on both the number and the structure of US universities. Building on the foundations established by the early research orientation of Johns Hopkins University and the expansion of academic medicine, as initiated by the Flexner report (Flexner 1910), the influx of federal support for research helped to change American universities into research-intensive institutions. For example, training was seen as part of the mission of NCI from its beginnings in the 1930s. Recodification of the Ransdell Act during 1944 reauthorized the training activities specified in the act. The training of scientists at the master's and PhD levels became an integral part of research. As new national institutes came into being, the authority for training–research or clinical–was often included as an essential component of their missions and incorporated into their statutory portfolio, as specified in Title IV of the Public Health Service Act. Funds to support the tuition and stipends of students and fellows were now often included as items in the budgets of federal research grants. By the early 1950s, NIH had administratively crafted an elaborate set of training mechanisms, including grants for predoctoral, postdoctoral, and special fellowships and for predoctoral and postdoctoral training; these supported a wide variety of training programs in the biomedical sciences.

The most general and comprehensive statutory authority for supporting research training was added to Section 301(d) of Title III of the Public Health Service Act by an amendment enacted in 1962 as part of PL 87-838. The amendment extended the limited authority of the surgeon general (later the secretary) from supporting simply "such research projects as are approved by the National Advisory Health Council" to supporting "such research and research training projects as are approved . . ." By the early 1970s, more than 6,000 life-science graduate students were supported by NIH and NSF training grants or fellowships. The National Research Act of 1974 (PL 93-348) established the National Research Service Awards program, providing funds for competitive individual fellowships for graduate students and postgraduate fellows. It also instituted a mechanism by which a committee appointed by the National Academy of Sciences met every 2 years to identify current national research training needs (NRC 1994). The new mechanism led to the termination of some training grants, but the general level of support for biomedical training continued to grow. The sums spent for life-science research training continued to mirror those spent for life-sciences research, as exemplified by the transient drop in the number of PhDs granted per year during the middle to late 1970s, which followed a temporary cessation in the rapid growth of research funding that occurred during the late 1960s. When federal research investments resumed growth in the middle 1970s, the rate of PhD production followed suit. The expansion of training has continued at various rates ever since, as detailed in chapter 2.

The growth of the life sciences has permitted the absorption into the research workforce of a large fraction of the ever-increasing trainees. The ready availability of recent PhDs has also contributed to the success of companies built on the life sciences, such as in the biotechnology industry. Scientists needed to guide company decisions and workers to staff research laboratories were already available when the discoveries of recombinant DNA in the 1970s empowered entrepreneurial scientists to develop processes that would make marketable products of an unprecedented kind. Human proteins could now be synthesized in large quantities outside the human body and used as therapeutic agents of great practical utility. During the 1980s, this industry grew rapidly, fueled in part by the enthusiasm of Wall Street for the possibilities associated with new markets. New investment from the private sector flowed quickly into the life-science enterprise, increasing both the quantity of scientific research and the perception that such work could be of value to the American people. In 1996, the number of life-science PhDs granted was 7,696; in 1997, federal investment in health research exceeded $14 billion. Private foundations contributed $1.2 billion to biomedical research in 1997, and industry's investment in health research and development exceeded $17 billion (NSF 1996, appendix table 4-31). Meanwhile, the country's investments in plant science and agriculture had also grown: during 1995, USDA invested $1.4 billion in research and development, and industry's investment in agriculture and forestry was $3.5 billion. The life-science research enterprise had become economically important.

In the recent decades, the various sectors of employment for life scientists have expanded at different rates. The fastest growth has occurred in industry, where the number of life-science PhDs has increased from around 5,500 in 1973 to nearly 24,000 in 1995, an average annual increase of almost 7%. During the same period, the pool of postdoctoral fellows and non-tenure-track staff at academic institutions has grown from about 4,000 to over 20,500, an average annual increase of 7.6%. In contrast, federal-laboratory and other government employment has shown modest growth; and the number of life scientists holding faculty appointments in universities and colleges has increased from 28,500 in 1973 to only about 49,000 in 1995, an average annual increase of only 2.5%. Universities remain the largest employers of life-science PhDs, but their share of the pool has diminished substantially during the last two decades (see appendix table F.8 for details).

Our country's investment in the life sciences has produced many important results. Discoveries in agricultural science have improved our understanding of soils and their chemistry and have led to the development of new strains of crop plants that are resistant to diseases and that yield more food per cultivated acre. Such work has contributed to the low cost of food that our country now enjoys. Environmental sciences and forestry have evolved new methods for sustainably managing resources that will help our expanding population to pass on more of its natural wealth to future generations. Medical science has provided fundamental understanding of the molecular basis of numerous diseases, which has led to the elimination of some and the containment of many. Not only preventive approaches, like proper nutrition and immunization, but diagnostic techniques and ameliorative treatments–drugs, surgery, radiation, and physical therapy and psychotherapy–have reduced human suffering and prolonged and enriched human life. Advances in molecular biology not only have spawned the biotechnology industry, which is contributing to the American economy, but also have contributed fundamental knowledge about the structures of genes and the behavior of biologic macromolecules. These advances are yielding new insights into the relationships among organisms and about the continuum of structure and function that connects living and nonliving things. (For more specific examples of the fruits of progress in the life sciences, see chapter 4.) The long-range implications of all this rapidly evolving knowledge are hard to predict, but many additional benefits are now on the horizon.

THE STRUCTURE OF THE LIFE-SCIENCE ENTERPRISE

The spectacular successes of the life sciences have emerged from a professional structure that evolved to meet the needs stemming from rapid growth. The lives of professors, industrial biologists, agricultural and medical researchers, postdoctoral fellows, and graduate students in the

1990s very different from those of comparable scientists 30-40 years ago. A typical academic research laboratory in earlier times included a professor, perhaps a technician, and sometimes a graduate student. Today, many life-science laboratories include 20 or more people, most of whom are in the process of training to become independent scientists. The chapters that follow present data on many aspects of the changes. To make the later chapters more meaningful for readers who are not themselves life scientists, we describe here the training of a life scientist and the major professional events in a life scientist's career–the work toward a PhD, in many cases postdoctoral training, the passage to a job, and the pursuit of research support–and then sketch the research environment. Space limitations require that this treatment be brief, so it is restricted in scope and detail; the descriptions are intended not to be detailed, but to illustrate what it is like to be trained and to work in today's biologic research enterprise.

It is important first to recognize the breadth of knowledge that is now encompassed by the term life sciences. At one extreme, we find physical and chemical studies of the molecules that make up living things: organic molecules–such as fats, carbohydrates, and proteins–that are the stuff of which all living things are made. The life sciences then range up through the study of genes and of the DNA and RNA from which they are constructed and expressed to studies of macromolecular assemblies and organelles and the cellular processes that they accomplish. Cells are sometime studied as organisms in their own right (for example, bacteria, protozoa, and some fungi) and sometimes as components of multicellular plants or animals, which must in turn be analyzed not only as organisms, but also as entities that develop from a single fertilized ovum and must interact with other plants and animals in their environments. Whole systems of interacting organisms must be studied to understand an ecologic niche. And the evolutionist would argue that none of the above studies makes sense unless viewed in the context of the slow changes in genetic makeup that constitute biotic evolution. All those aspects of the life sciences are linked by the universality of the genetic and biochemical bases that underlie them, but it is clear that there are many ways to study the complexities of life.

The life sciences can be thought of in three categories: the agricultural sciences, the biomedical sciences, and a harder-to-label cluster of basic biologic sciences that address life processes themselves. This report includes data from all those categories, and we have tried to address the interests of every federal agency that supports training and research in biology, broadly defined. It might appear at times that NIH and the biomedical sciences have dominated our considerations. That appearance has been difficult to avoid because of the size of the NIH budget and the resulting number of young and established life scientists that it supports. Indeed, patterns of support that are initiated by NIH often serve as models for programs funded by other agencies. We hope that our discussions and recommendations will be relevant to all the life sciences, not simply those with a biomedical bent.

THE SHAPE OF GRADUATE EDUCATION

All new graduate students in biology must select from a panorama of topics, like that sketched above, a specific subset that can reasonably be mastered within the 5-10 years that are commonly devoted to a PhD degree. Graduate work almost always begins with courses, but many programs strive to get their students into a research environment as soon as possible. The intent is partly to distinguish graduate from undergraduate education and partly to let students see what the life of a scientist is like. Coursework usually dominates the first year or more of graduate study and trickles on through years 2 and 3. A preliminary examination usually evaluates competence to continue training, and the passage of a general examination in the second or third year permits admission to candidacy for the PhD degree. A graduate student usually identifies dissertation supervisor in the first or second year and begins thesis research shortly thereafter.

It is uncommon for graduate biology students to pay their educational expenses from their own resources (see table 2.1 in chapter 2), because there are numerous alternatives: salary grants to individual students, training grants to departments or programs, research grants to faculty members who can then support a graduate research assistant, teaching assistantships from the college or university, and in some cases loans to help to postpone expenditures until more lucrative employment is available. Most graduate students teach at some time during their training, but the duration of this teaching experience usually depends on whether they can obtain support from a research-oriented source that allows them to complete their thesis work without the complications of teaching at the same time.

The duration of graduate training is variable, depending in part on the subdiscipline in question: molecular biology and cellular biology tend toward 7 years (elapsed calendar time from the bachelor's degree to the PhD degree and about a year less as a registered student in the program), but training that requires extensive work in the field or an analysis of populations over a long term takes longer. The mean time to completion of a life-science PhD has increased from 6 to 8 years over the last 25 years. (Chapter 2 presents more detailed data on the graduate and the postdoctoral experience.)

THE POSTDOCTORAL EXPERIENCE

Graduate students in biology who desire a career in research often pursue further training at the postdoctoral level. According to data from the National Research Council's Survey of Doctorate Recipients (SDR, see, for example, NRC 1996), the fraction who go on to this level of training more than quadrupled from 1973 to 1993; in 1995, 53% of life-science PhD recipients pursued further training as postdoctoral fellows within 1-2 years of earning their degrees. Three reasons for postdoctoral training's becoming so common in the life sciences have been suggested: building a successful research career requires such a magnitude and diversity of knowledge that additional training in a second research environment is helpful; funds are often available for postdoctoral stipends, making the second training stage relatively available and additional outlays by the postdoctoral fellow unnecessary; and the competition for jobs with more independence and security is intense. Thus, the improvements in one's curriculum vitae (CV) that result from the additional research experience and publications characteristic of postdoctoral work are very important for one's prospect of permanent employment. The relative importance of those factors is discussed in chapter 5.

Some postdoctoral fellows apply for and receive their own funding from a government agency or a private foundation. Such fellowships are particularly desirable because the recognition that accompanies them carries implicit and explicit messages of intellectual and professional independence and because the salary money makes a candidate more attractive to a host laboratory of high quality. Other postdoctoral fellows are supported by salaries specified in the research budgets of their new host laboratory. To some extent, scientists in the latter group are more obliged to work on the projects for which their new mentors have been funded than on projects of their own choosing. However, because postdoctoral fellows commonly select their host laboratories on the basis of an interest in the science that is done there, that constraint is usually of minor importance, at least at first. Many young scientists find that the first 2 or 3 years of postdoctoral experience is exceptionally rewarding. Researchers at this stage of professional development are already experienced enough to get good work done fast, but new enough to the subdiscipline of their new host laboratories to find their work both challenging and valuable. The combination of scientific competence with a new scientific project is heady, constructive, and useful. Many senior scientists look back on their postdoctoral years as among the best of their scientific careers.

The graduate experience and postdoctoral training are formative in developing a sense of how science should be done. Virtually all graduate training and most postdoctoral work are carried out in the academic environment of a university or medical school, so the experiences of young life scientists are heavily weighted toward the loosely structured environments characteristic of basic-research laboratories. That situation might contribute to the preference that many postdoctoral fellows show for continuing their careers in an academic environment.

In recent years, it has become common for postdoctoral training to last at least 3 years. That situation is now having an important on the lives of older postdoctoral fellows because most postdoctoral fellowships last for only 2 or 3 years. For those who derive their stipends from host laboratories or institutions, the support rarely extends more than 5 years. A distinction should be made between "postdoctoral training", when a young life scientist is learning new approaches or techniques, and "postdoctoral employment", when training is largely over and the young scientist is continuing to work at this professional rank, improving his or her CV and/or looking for a more permanent and independent job.

As the length of the postdoctoral experience increases, the issue of job security can become more important. Moreover, starting postdoctoral salaries are usually rather low and increase only modestly with additional years of experience (the recommended NIH postdoctoral salaries for a person with up to 5 years of previous postdoctoral experience have recently been increased to just over $20,000 per year at the beginning of their NIH-supported postdoctoral work and just under $30,000 per year at the end; fringe benefits are also modest). Few universities have a professional structure that provides additional financial support for postdoctoral fellows, and although they are welcomed in scientific professional societies, they are neither students nor established professionals. That situation provides strong motivation for most postdoctoral fellows to try to find a different form of employment within 5 years of obtaining their PhD degrees.

THE PURSUIT OF A JOB

After a period of postdoctoral training and the publication of several papers as evidence of scientific accomplishment and expertise, most postdoctoral fellows apply for positions that carry some measure of future prospects and permanence: tenure-track academic posts, jobs in companies or government laboratories, or positions in alternative professions that will enable them to use their scientific training or research skills. In recent years, the job market for life-science PhDs has tightened considerably. The number of positions in academic institutions, the largest employers of life-science PhDs, has not increased as fast as the number of applicants. Junior faculty positions for which the field of research is not narrowly defined generally attract several hundred applicants, and good jobs in industry and in primarily undergraduate, teaching-intensive colleges are just as competitive. Of course, some young scientists with extraordinary credentials get jobs immediately, but many others with impressive CVs are now finding the professional transition very difficult (for a more complete treatment of this important issue, see chapters 2 and 3).

In response to the tightening job market, there has been an expansion in the range of positions that young biologists will look at seriously. The extent of this "alternative" job market is not at present very clear, but some of the major research centers are beginning to provide symposiums and conferences on the careers available to life-science PhDs outside the conventional spheres of employment. The reaction among postdoctoral fellows has been mixed (as discussed in chapter 5). The problem for an individual postdoctoral fellow remains how best to be distinguished from the competition. To maximize their marketability, most candidates try to publish as much as they can in journals that are widely read. Job seminars get brightly polished, and candidates practice presenting themselves favorably. Even with strong credentials and a broad perspective on the suitability of diverse employment opportunities, however, it often takes several years to get a good job. This difficulty is almost certainly an important factor in the increasing duration of postdoctoral "training".

THE PURSUIT OF RESEARCH SUPPORT

For applicants who get positions in industrial or governmental laboratories, resources for research are usually included. For new employees in academic institutions and research institutes, the next career step is usually to obtain funding that will support scientific work. Many job offers include some funds with which to set up laboratories, so initial purchases of equipment and often the first year or so of research supplies are already available, but the expectation for most new employees in these research institutions is that they will apply for and obtain their own research funding. The details of an application vary from one granting agency to another, but a research proposal usually includes a description of the scientific context and significance of the proposed experiments and a detailed account of how the work will be done. Construction of such a proposal takes anywhere from a few weeks to a few months, and the probability of success of first applications is not high, ranging from less than 10% in some agencies to 35% in others. Such figures, of course, vary from year to year and depend primarily on the state of the economy and the attitude of Congress toward research.

Staying funded is not much easier. It is important to remember that obtaining grants has been difficult for many years; there are few investigators still submitting proposals whose work is not of good quality. The competition is therefore intense for all investigators, young and old, and achieving a rank in the top one-third is not easy. A successful proposal requires not only imagination, skill, and hard work, but also good fortune. It helps to be in the right intellectual place at the right scientific time. If a proposal is radically different from the scientific mainstream, it can be dismissed as "risky". If it is not sufficiently involved with current methods and ideas, it can be dismissed as old-fashioned. There is also some luck in the rather arbitrary choice of who reviews a particular proposal. Most reviewers are highly accomplished scientists, chosen by well-meaning grant administrators for their expertise and fair-mindedness. However, when the people who review a proposal know and respect both the subfield in question and the work of the applicant, the chances of a fundable score are likely to improve.

It is also important to recognize the importance of funding for life scientists working outside government or industrial laboratories. Most universities, medical schools, and research institutes require grants to individuals for the pursuit of a particular project: if there is no grant, there is no (or very little) support for research. Furthermore, one's livelihood is often affected by a grant, dramatically in some instances. In most colleges of arts and sciences and related university divisions, a salary is provided for only 9 months of the year, the time when a principal investigator is engaged in teaching and related university activities. Salary for the summer months can be sought from a research grant, and sometimes a fraction of a principal investigator's academic-year salary will be included as well, on the grounds that the faculty member is using that portion of his or her time on research-related activities. In medical schools and other medical research institutions and in private institutions to a greater extent than in public ones, research personnel are expected to obtain substantial portions of their salaries from grants throughout the year. Thus, the motivation to write successful proposals is high indeed. Given all those factors, it is no wonder that many principal investigators spend a large fraction of their time seeking the funds with which to do research.

THE CHARACTER OF THE RESEARCH ENVIRONMENT

Given the diversity of biologic research, there is a huge range in how life-science research is conducted. Some is done "in the field", with a heavy emphasis on the observation of organisms in their natural settings. Some is done in the field, literally; selected plants are grown in experimental plots side by side with control strains to assess their relative susceptibility to disease, drought, or nutritional deprivation. Some is done in laboratories that could serve a chemist or a physicist as well as a biologist. The following generalizations should, however, be reasonably applicable to all.

A principal investigator builds a research group by defining the scientific questions to be addressed, specifying the methods to be used, obtaining necessary funding, finding the suitable research environment, and attracting the research personnel, usually a mixture of students, technicians, and postdoctoral fellows. The day-to-day jobs of the principal investigator include those of a research manager: making decisions about expenditures and personnel matters, evaluating data, planning the next experiments or observations, providing training for less experienced personnel, and directing the whole enterprise toward the completion of research manuscripts for publication. Ancillary tasks include the writing of grant proposals and such research-related articles as reviews of the literature, critiques of work of other principal investigators, and the committee work associated with the host institution. Many principal investigators must also teach and administer activities distinct from their own research projects.

The research personnel in the group usually work on more-specific tasks that pertain to the construction of research tools or the acquisition and analysis of data. Group sizes usually range from a few workers to around 20; some exceptional research groups are much larger. It is common for the social structure of the research environment to be quite free, permitting and even encouraging iconoclastic and innovative contributions from anyone in the group. Rarely is the judgment of the principal investigator always right, and the details of a particular experiment or observation are sometimes known only to the people doing the work. The ebb and flow of criticism and suggestion between the principal investigator and the laboratory members is one of the things that make a free social structure so effective for the progress of science. The give and take is one of the most instructive and constructive aspects of a laboratory environment; it is a key reason why research training must be obtained "on the job" in an apprentice situation, not in a classroom. The give and take is also of great value for the quality and quantity of science that gets done; mistakes in judgment or knowledge are often corrected quickly without the emotional stress that can develop in a more structured environment. It is the rare (and foolish) principal investigator who is offended by constructive disagreement.

One of the most important aspects of the laboratory group structure is its flexibility and intellectual mobility. In fast-moving fields like the modern life sciences, the intellectual ossification that can accompany a major administrative structure, such as the environment suitable for an expensive instrument, impedes the readjustments of position and direction that are necessary for innovative work. Flexibility of structure has been one of the great strengths of life-science research in the United States. Research groups can vary widely from the model described above, depending on the discipline, the size of the group, the personality of the individuals involved, and the institution; but even this variation is probably constructives: it allows the country's research enterprise to encompass many approaches within the framework of research that is supported by grants to individual life-science investigators. The resulting pluralism has contributed to the ability of American life-sciences to explore the biologic landscape fast and economically. Even the research structures found in many companies can be described by this model, although they include a different range of constraints, depending on the scientific and economic goals of the companies.

REFERENCES

Flexner A. 1910. Medical education in the United States and Canada. New York: Carnegie Foundation.

Judson HF. 1979. The eighth day of creation: makers of the revolution in biology. New York: Simon and Schuster.

NIH (National Institutes of Health). 1961. Basic data relating to the National Institutes of Health. Bethesda, MD: NIH.

NIH (National Institutes of Health). 1978. NIH almanac. Bethesda, MD: NIH.

NRC (National Research Council). 1994. Meeting the nation's needs for biomedical and behavioral scientists. Washington, DC: National Academy Press.

NRC (National Research Council). 1996. Summary report 1996: Doctorate recipients from United States universities. Washington, DC: National Academy Press.

NSF (National Science Foundation. 1960. Science, the endless frontier, a report to the president on a program for postwar scientific research. Washington, DC: US Government Printing Office [Reprint of 1945 publication.]

NSF (National Science Foundation). 1996. Science & Engineering Indicators 1996. NSB 96-21. Washington DC: US Government Printing Office.

2 EDUCATION AND RESEARCH TRAINING OF LIFE-SCIENCE PHDS

In this chapter, we examine the changes that have occurred over the last 3 decades in the number of new life-science PhDs produced and the length of their doctoral and postdoctoral training. We also examine some key factors underlying these trends to establish a basis for understanding the forces that influence the trends in career outcomes presented in chapter 3. Most of the data in this chapter come from two National Research Council surveys: the annual Survey of Earned Degrees, which collects biographic information (including postdoctoral plans) from all persons receiving research doctorates from US universities, and the biennial Survey of Doctorate Recipients, which compiles current employment information from a 5-10% sample of US-educated PhD scientists and engineers. Additional data on graduate-student support and postdoctorals were obtained from the National Science Foundation's Survey of Graduate Students and Post-doctorates in Science and Engineering. (See appendix C for additional detail on sources of data and appendix D for fields of study included in the committee's analysis.)

PHDS AWARDED IN THE LIFE SCIENCES

Since the 1960s, the number of PhDs awarded annually in the life sciences has more than tripled. As illustrated in figure 2.1, 7,696 life-science doctorates were awarded by US universities in 1996, compared with 2,095 degrees in 1963. However, the growth pattern during that 33-year period has not been constant. During the first 8 years, primarily as the result of the many new graduate programs that were established[1] and programs that were expanded before 1963 (as discussed in chapter 1), the number of PhD awards grew at an average of 11.4% a year. In the next 16 years (1971-1987), there was minimal growth in PhD production (less than 1% a year). Since 1987, the growth in doctoral degrees in the life sciences has resumed–an average of about 4% from 1987 to 1996 (the most recent year for which data are available), for a total increase of 42.5% in that period. (See table E.1, in appendix E, for details and figure 2.1 for a graphic presentation.)

The increases in PhD awards have by no means been uniform across the disciplines of the life sciences. Changes in survey taxonomy do not permit a detailed analysis of the doctoral increase in every life-science discipline, but some of the differences observed from data in tables E.2, E.3, and E.4 are striking. For the most part, the largest increases have occurred in biomedical sciences, such as biochemistry, cellular biology, molecular biology, neurosciences, and pharmacology. The numbers of PhDs awarded in some agricultural and basic biologic sciences (such as plant sciences and ecology) have also grown during the last 3 decades but to a much smaller degree. Overall, almost all the growth in the number of PhDs awarded has been in the biomedical fields (figure 2.2).

Two demographic characteristics of life scientists have changed considerably during the 30-year period under study. First, as can be seen in figure 2.1, the percentage of PhDs awarded to women has grown steadily. In 1963,

---

[1] Between the late 1950s and 1970, the number of PhD-granting programs in the life sciences grew from 122 to 224 (NRC 1978)

Figure 2.1 Number of US life-science PhDs awarded annually, by sex, 1963-1996.

Data from table E.1. 1996 total includes five recipients of unknown sex.

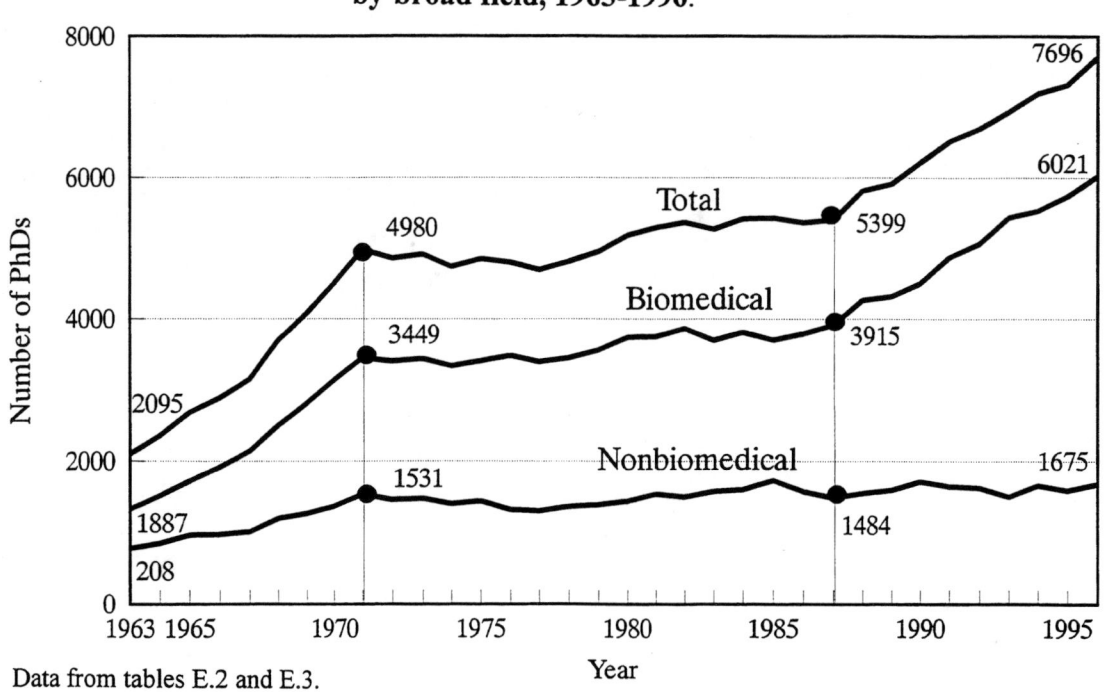

Figure 2.2 Number of US life-science PhDs awarded annually, by broad field, 1963-1996.

Data from tables E.2 and E.3.

for example, fewer than 10% of life scientists receiving PhDs were women. By 1996, the corresponding fraction was over 40%. In contrast, the number of men receiving life-science PhDs–after rapidly rising in the 1960s–actually declined from 1971 to 1987 and has only modestly increased since then. Although there has been a doubling in the fraction of life-science PhD recipients who are members of minority group over the last 20 years (table E.1), the absolute numbers remain very small–rising from 96 in 1973 to 341 in 1996.

The second notable change is the increase since 1987 in the number of degrees awarded to citizens of other countries. As shown in figure 2.3 and table E.5, the number of foreign citizens (holding permanent-resident status or temporary visas) earning life-science degrees at US universities more than doubled from 1987 to 1996 (from 1,127 to 2,947). The percentage of life-science PhDs who are foreign nationals with temporary visas peaked at 28.2% in 1993 but declined somewhat thereafter. That is almost certainly an artifact attributable to the passage of the Chinese Student Protection Act of 1992, which permitted Chinese nationals temporarily residing in the United States to change to permanent-resident status; many Chinese students who have earned PhDs since 1992 have been counted in the US citizen and permanent-resident category. Figure 2.3 shows that when the number of temporary residents receiving PhDs dipped after 1993, the number of permanent residents increased sharply and that the sum of these two classes of foreign nationals rose at a steady pace from 1989 to 1996.

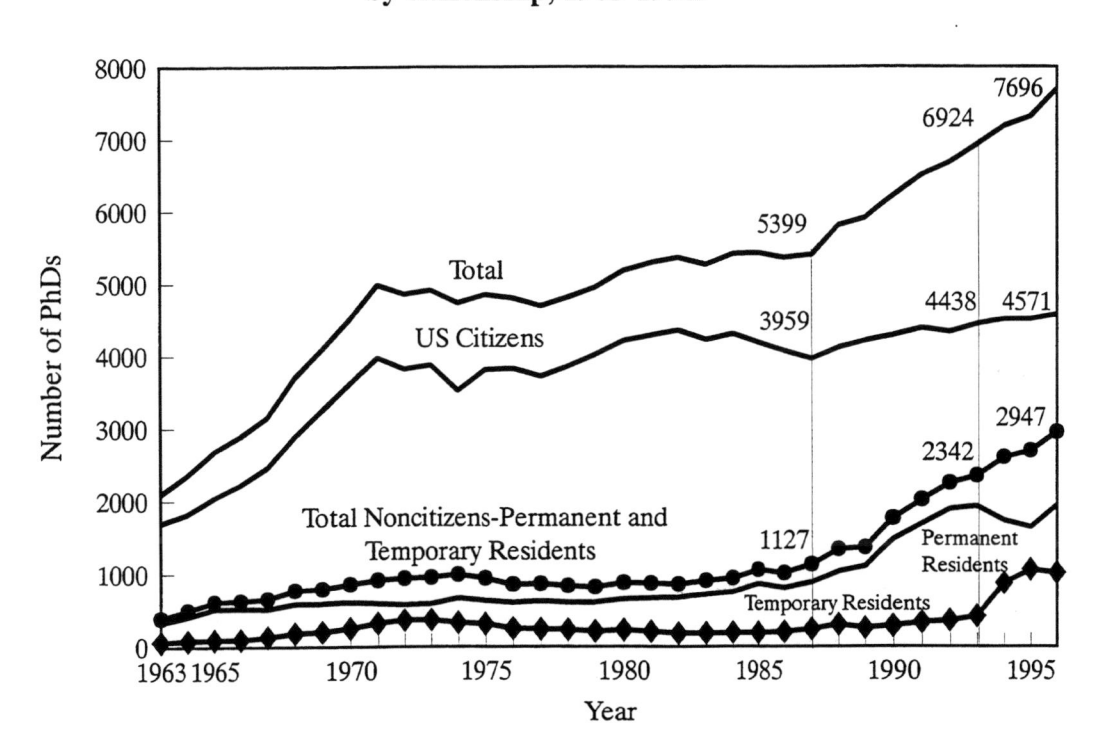

Figure 2.3 Number of US life-science PhDs awarded annually, by citizenship, 1963-1996.

Data from table E.5. 1996 total includes 178 recipients of unknown citizenship.

We do not have accurate data on how many of the foreign students on temporary visas have pursued research careers in the United States, but the percentage appears to be substantial. Figure 2.4 shows that an increasing number and percentage of temporary residents report on receiving their PhDs that they plan to remain in the United States. In recent years, about 60% have said that they plan to remain. Finn and others (1996) estimated that nearly one-third of the temporary residents who earned life-science PhDs in 1987-1988 were working in this country in 1992. The foreign-national PhDs are found in the highest proportions in subdisciplines of the agricultural sciences–such as agronomy, animal breeding, food engineering, and plant pathology– and fields that have more direct application, such as pharmacy.

Figure 2.4 Number of US life-science PhDs awarded annually to temporary residents and number and percentage of temporary residents planning to remain in the United States, 1963-1996

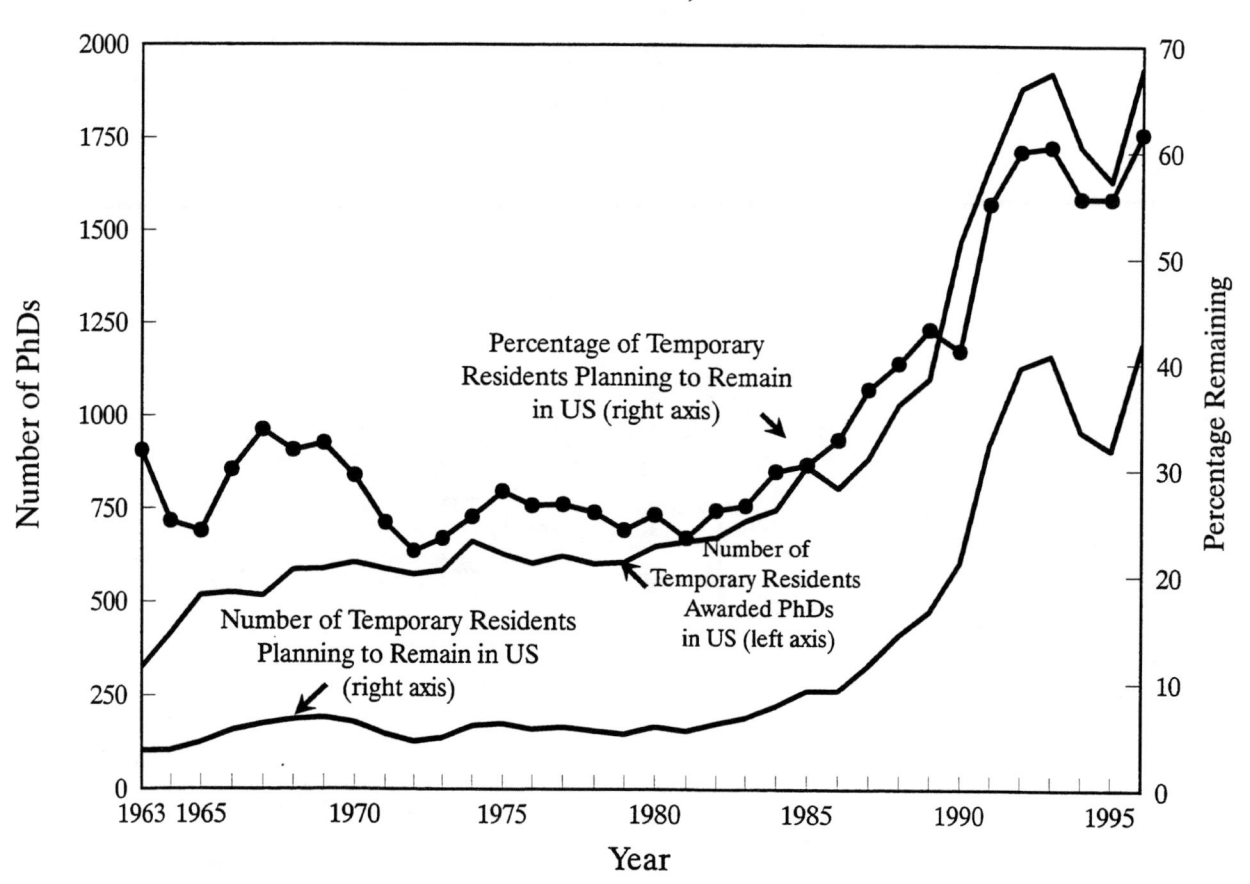

Data from table E.1.

Although women and foreign nationals account for most of the increase in the number of PhD recipients over the last 10 years, there is a notable difference in the academic standing of the institutions in which they train. Overall, the top 26 life-science PhD-granting programs by reputation[2] (NRC 1995) educate 25-32% of the life-science PhDs, a percentage that has remained roughly constant over the last 3 decades. Their programs have historically awarded a disproportionate share of the doctorates received by women. For example, in 1963, when the top programs granted 34% of all PhDs, they awarded 45% of all PhDs going to women. Although the percentage has fallen, it consistently has stayed above the top 26 programs' share of total degrees awarded; women who receive PhDs are consistently more likely to get their degrees from top departments than are men.

In contrast, the large increase in the proportion of degrees awarded to temporary residents occurred primarily at non-top-26 institutions. Only in the very earliest years was it as high as, (or higher than) the proportion of all degrees awarded by the top 26 programs, and during most of the period it was substantially lower.

---

[2] In alphabetical order, the top 26 institutions are: Baylor College of Medicine, Brandeis University, California Institute of Technology, Columbia University, Main Division, Duke University, Harvard University, Johns Hopkins University, Massachusetts Institute of Technology, Northwestern University, Princeton University, Rockefeller University, Stanford University, University of California, Berkeley, University of California, Davis, University of California, Los Angeles, University of California, San Diego, University of California, San Francisco, University of Chicago, University of Michigan, University of North Carolina, Chapel Hill, University of Pennsylvania, University of Texas/ Southwest Medical Center, University of Washington, University of Wisconsin, Madison, Washington University, and Yale University. The list includes 26 institutions because there was a tie for 25th place.

There has also been a change in the means of financial support of graduate students–an increase in the fraction of graduate students receiving federal and institutional support and a large increase in the fraction supported as research assistants. As shown in figure 2.5 and table 2.1, the fraction of life-science graduate students receiving federal funds rose from 28.3% in 1975 to 28.7% in 1985 and to 34.8% in 1995. Almost all the increase between 1985 and 1995 is attributable to the support of students by research grants; the fraction of students supported by federal training grants or fellowships fell during the same period. The number of students supported by institutional (university) funds increased markedly, almost entirely because of the larger number supported as research assistants. The relatively small fraction of self-supported students dropped sharply between 1975 and 1985.

Table 2.1 is a snapshot in time of the primary means of support. In the course of a graduate student's education, the student might shift from one means of support to another. Data show that about two-thirds of students receive federal support at some time in their training

TIME REQUIRED TO ATTAIN THE PHD

The time required to complete the PhD in the life sciences has increased substantially over the last 30 years. As illustrated in figure 2.6, the median time to finish requirements for the doctorate–as measured from graduate enrollment to PhD award (that is, total elapsed time)–has increased from 6.0 years for 1970 graduates to 8.0 years for 1995 graduates. As can be seen from the data presented in table E.4, this median time has varied considerably among disciplines. For example, fields that either involve extensive field work–such as ecology, forestry, conservation, and fish sciences–or require multiyear studies–such as epidemiology and public health–have typically

Figure 2.5 Primary source of support of graduate students in life sciences, 1975, 1985, 1994.

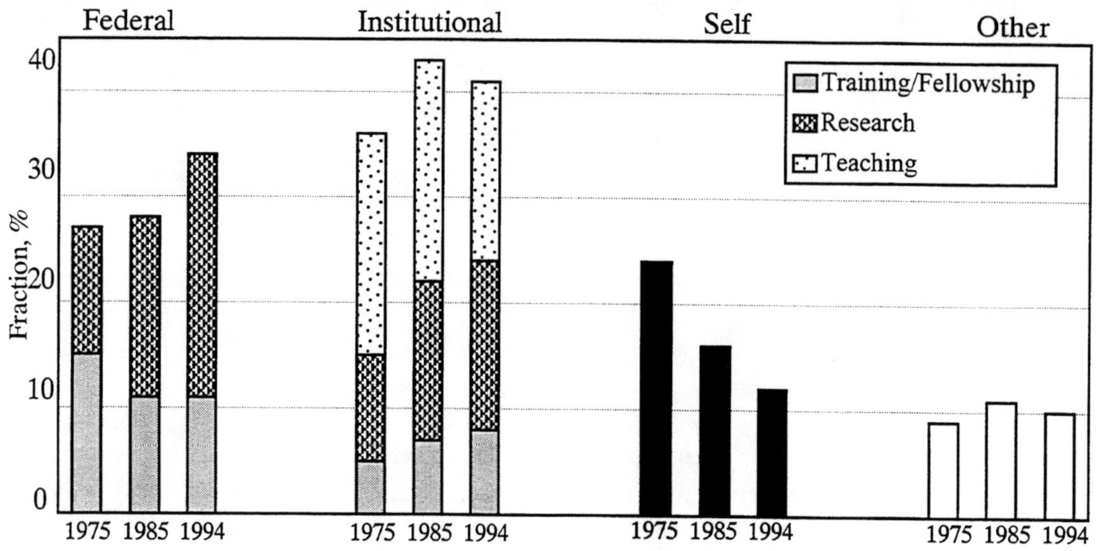

Data from table 2.1.
Data not available on function for self supported and other supported.

Table 2.1 Number and Percentage of Graduate Students of Various Kinds and Sources of Support, 1975, 1985, 1995

	1975 No.	1975 % of Group	1975 % of Total	1985 No.	1985 % of Group	1985 % of Total	1995 No.	1995 % of Group	1995 % of Total
Federal support									
Research assistant	4653	41.7	--	6928	58.6	--	11963	66.5	--
Trainee/fellow	5994	53.7	--	4285	36.2	--	5391	30.0	--
Teaching assistant	118	1.1	--	96	0.8	--	155	0.9	--
Other	404	3.6	--	512	4.3	--	471	2.6	--
Total federal	11169	100.1	28.3	11821	99.9	28.7	17980	100.0	34.8
Institutional support									
Research assistant	3876	25.3	--	5678	31.2	--	8489	38.2	--
Trainee/fellow	2040	13.3	--	2891	15.9	--	4017	18.1	--
Teaching assistant	8495	55.5	--	8647	47.5	--	8589	38.6	--
Other	901	5.9	--	978	5.4	--	1136	5.1	--
Total Institutional	15312	100.0	38.7	18194	100.0	44.2	22231	100.0	43.0
Other									
Self-supported	9359	71.8	--	6388	57.2	--	6396	55.5	--

Education and Reseach Training of Life-Science PhDs

	1975			1985			1995		
	No.	% of Group	% of Total	No.	% of Group	% of Total	No.	% of Group	% of Total
(cont'd)									
Private and foreign	3676	28.2	--	4786	42.8	--	5124	44.5	--
Total other	13035	100.0	33.0	11174	100.0	27.1	11520	100.0	22.3
GRAND TOTAL	39516	--	100.0	41189	--	100.0	51731	--	100.1

Source: NSF 1995.

had longer doctoral training periods than disciplines that focus on laboratory-based research. Nevertheless, in every life-science discipline, the median time to complete the PhD is longer now than it was 2 decades ago. Since 1992, there has been no increase in median time to degree.

Figure 2.6 Median elapsed time to PhD and age at time of PhD, 1970-1996.

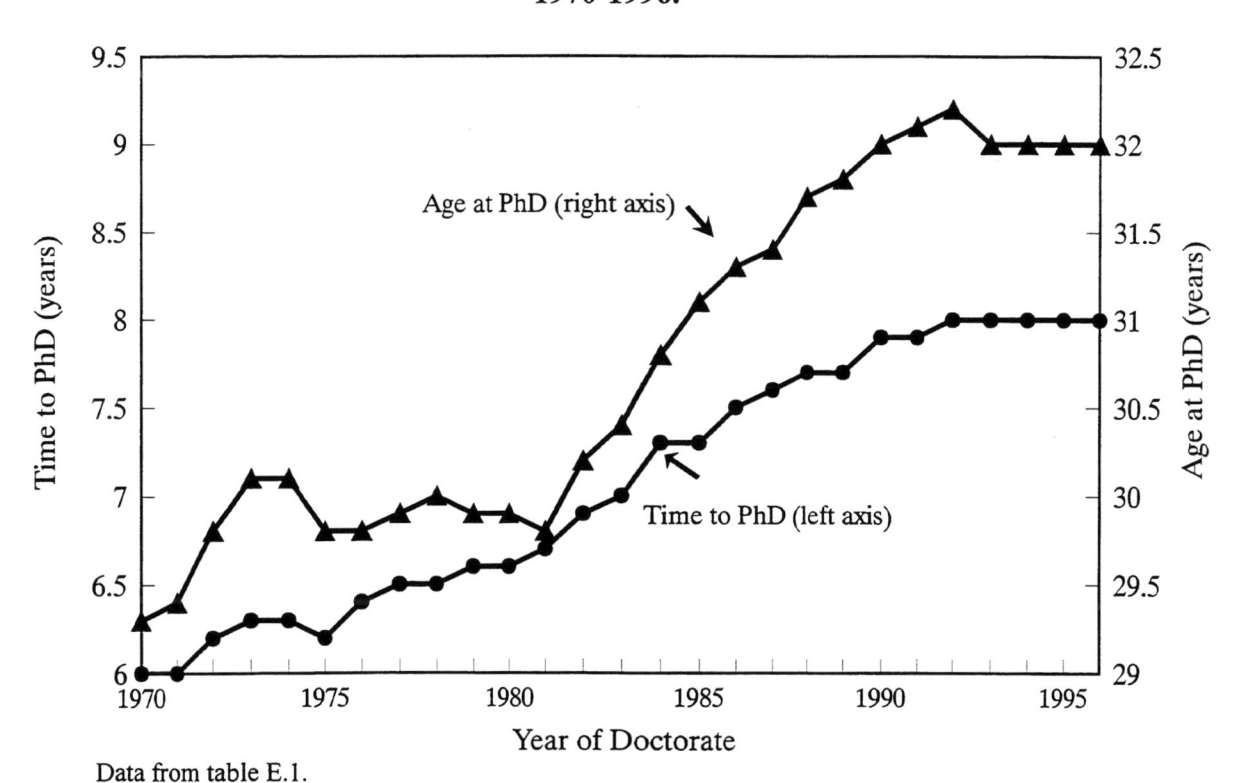

Data from table E.1.

Not unexpectedly, recent PhD recipients are completing their degree requirements at higher ages than their colleagues who graduated in the 1970s and 1980s. The data in figure 2.6 reveal that the median age at PhD has risen from 29.3 years for 1970 graduates to 32.0 years for 1996 graduates. This increase of 2.7 years is substantially greater than the increase of 2.0 years in median time to complete graduate training. The difference might be explained by that fact that students have been enrolling in graduate programs at higher ages–especially in recent years.

It is uncertain why the time to degree has lengthened. No compelling academic reason exists, inasmuch as coursework typically is completed within 2 years and research usually begins at the end of the first year. Some argue that faculty use graduate students as a source of labor to conduct faculty members' research. Others point to possible benefits for the students, such as an opportunity to increase the numbers of publications on which their names appear. Without a cap on the number of years of support, there might be no compelling reason to complete the degree, especially given the perceived unfavorable job market. Students could also be trying to wait out a period of poor employment possibilities by stretching their time in school and building their resumes. It should be noted that there has been no increase in elapsed time to degree or age at degree after 1992.

POSTDOCTORAL TRAINING

For a steadily increasing fraction of life-science PhDs, receipt of the doctoral degree does not signify the completion of research training. As shown in figure 2.7, both the number and the percentage of PhDs planning to take postdoctoral appointments after graduation rose dramatically from 1963 to 1992. From 1993 to 1996, the number of PhDs planning postdoctoral training increased, but the percentage decreased somewhat. In the middle 1960s, fewer than one-fourth of the life-science graduates planned postdoctoral work; by the late 1980s, the fraction had doubled. The trend resulted in an increase in the total number of graduates planning postdoctoral work from 485 in 1963 to 3,940 in 1996. As will be discussed in chapter 3, that phenomenon has had a dramatic effect on the career patterns of young life scientists.

Although the trend has occurred in all life-science disciplines, it should be emphasized that the likelihood of a degree recipient's taking a postdoctoral position has varied greatly from one field to another (see table E.4). In many of the agricultural sciences, for example, fewer than one-fourth of the recent (1986-1996) graduates have planned postdoctoral work; in some biomedical disciplines such as molecular biology and neurosciences, more than three-fourths of the PhD recipients have pursued additional research training.

Figure 2.8 shows the growth in the number of postdoctoral fellows (both US citizens and foreign nationals) in academic institutions, which has increased steadily since 1972. By 1995, the number of academic postdoctoral fellows had reached 15,348 (NSF 1995). In addition to the postdoctoral fellows in academe, there are postdoctoral fellows in government laboratories (about 3,200, including clinical fellows at the National Institutes of Health) and in industry. A 1995 survey by the American Society for Microbiology (Van Ryzin and others 1995) found that 763 PhD microbiologists in industry (11% of the 7,090 PhD microbiologists in industry) were postdoctoral fellows. We estimate the total population of postdoctoralss at about 20,000, but the number could well be higher.

Figure 2.7 Number and percentage of PhDs planning postdoctoral training on graduation, 1963-1996.

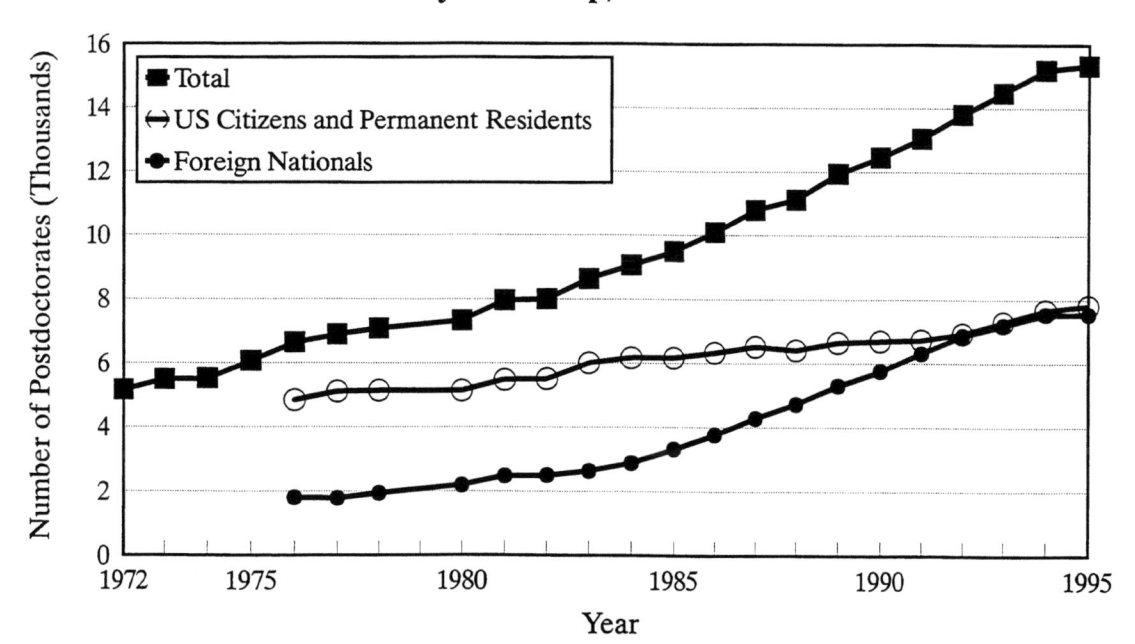

Data from table E.1.

Figure 2.8 Postdoctorates in biologic and agricultural sciences, by citizenship, 1972-1994.

Source: NSF/SRS Selected Data on Graduate Students and Postdoctorates in Science and Engineering, Fall 1993, Selected Data Tables, J. G. Hukenpohler, and NSF/SRS, 1995 Survey of Graduate Students and Postdoctorates in Science and Engineering.

Over the last 20 years, foreign nationals have made an increasing contribution to the size of the postdoctoral pool. In 1975, they held about one-fourth of all academic posts; in 1995, they held half the academic postdoctoral positions. In one important nonacademic environment–the National Institutes of Health intramural postdoctoral program–almost exactly half the postdoctoral work-ers are foreign citizens (Michael Fordis, National Institutes of Health, 1996 personal communica-tion).

It is important to understand that the data and discussions of chapter 3 and the remainder of this report generally do not include the large number of foreign citizens who, after completing their doctoral education abroad, have come to this country for postdoctoral training. Those scientists and foreign citizens who have obtained their PhDs here but declared their intention to leave the country are not included in the Survey of Doctorate Recipients, so there is no systematic evidence available to chart their career paths. Some data indicate that, at least in one sector, foreign nationals compete well for positions in this country. Association of American Medical Colleges (AAMC) data indicate that in the late 1980s and in the 1990s, close to one-third of new hires of PhDs, MD-PhDs, and MDs whose primary responsibility was research in basic-science departments were foreign nationals (special analysis for this study from AAMC Faculty Roster System by Lisa Sherman, 1997; see table E.9).

LENGTH OF POSTDOCTORAL TRAINING

From committee members' experience and from much anecdotal evidence collected by the committee, it appears that many postdoctoral fellows are spending longer times in training in recent years–4 or more years is now not uncommon for young biomedical scientists in some fields. The trends presented in figure 2.9 and table E.9, based on a retrospective reporting by respondents to the 1995 Survey of Doctorate Recipients, confirm the impression. The fraction of young life scientists holding postdoctoral appointments longer than 2 years increased substantially among those graduating in the late 1970s, with more modest growth since then. A similar pattern is observed for the fraction holding postdoctoral appointments for a total of more than 4 years. It is too early to obtain reliable estimates for graduates of the 1990s because some of them have not yet completed their postdoctoral work. Furthermore, figure 3.3 (in chapter 3) shows that a higher fraction of PhDs were in postdoctoral training in 1995 3-4 years and 5-6 years after they received their degrees than in 1973 and that the increase is greatest in the cohorts that received their degrees 3-4 and 5-6 years earlier. It is not possible to establish from these data a meaningful median time spent in postdoctoral work. However, there are clear indications that more young scientists are spending long periods as postdoctoral fellows.

On the basis of data and discussion above, it is evident that over the last 2 decades life-science PhDs have been spending increasing amounts of time preparing for research careers–a consequence mainly of the longer period in graduate training and the larger fraction that take post-doctoral fellowships of long durations. Most students pursuing a biomedical science career, for example, can now expect to spend 6 or more years in graduate school, and many spend another 4 years or more in postdoctoral work.

FIELD AND OTHER DIFFERENCES

This chapter frequently notes differences among sectors of the PhD population. The reader is referred to tables E.4 to note important

Education and Research Trainings of Life-Science PhDs

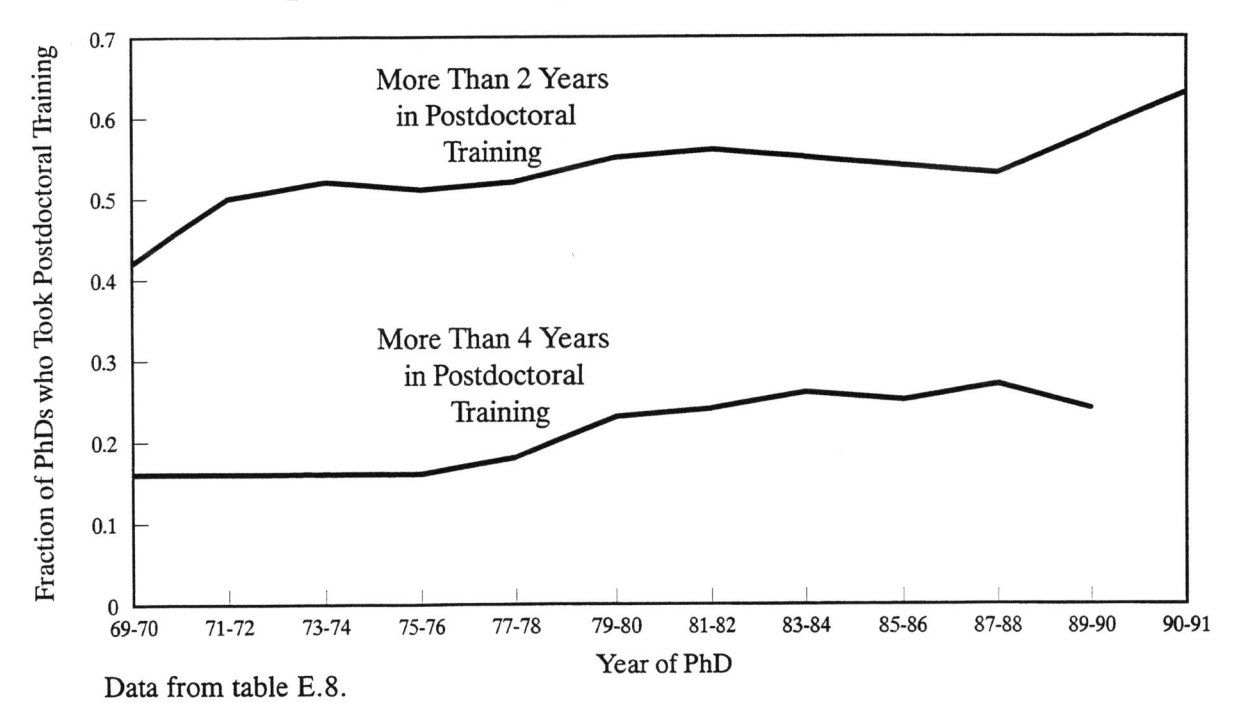

Figure 2.9 Time spent in postdoctoral training by life-science PhDs who took postdoctoral training, 1969-1994, as reported in 1995.

Data from table E.8.

differences, for example, that almost all the increase in the life-science PhD population is in biomedical sciences, whereas there has been little or no increase in the number of nonbiomedical-science PhDs. Table E.10 shows differences by sex, race, citizenship, and top-26 universities compared to non-top-26 institutions.

SUMMARY

- The number of life-science PhDs awarded annually in the United States has increased by 42% since the late 1980s, and the number awarded in 1996 was more than 3 times the number awarded in 1963.
- Foreign nationals with either permanent or temporary visas accounted for 38% of the life-science PhDs in 1996, and the number of temporary-visa holders planning to remain in the United States has risen to about 60% in recent years.
- Almost all the increase in numbers of life-science PhDs awarded has been in biomedical fields; the number in nonbiomedical fields has remained virtually the same since 1970.
- The median elapsed time between entry into graduate school and receipt of the life-science PhD has increased by about 2 years, from 6 to 8 years, but PhDs are obtained more quickly in some fields.
- The federal government financially supports the education and research training of about one-third of all life-science graduate students. The almost 12,000 graduate students supported by federal research grants represent the largest support mechanism among all categories of support–federal, institutional, or self.
- An increasing percentage of life-science PhDs do postdoctoral work, and the length of time spent in postdoctoral training is increasing.

- The number of persons in the post-doctoral pool has been increasing steadily and is now about 20,000.

Those several changes have had a serious effect on the labor market for life scientists. Throughout the roughly 30-year period being considered in this report, the cohort of young scientists entering the workforce has been much larger than the cohort that they replace (those who had completed their training 30 or so years earlier). Although the number of life scientists reaching retirement age has been steadily increasing, so has the number entering the workforce. For example, some 2,700 doctorates were awarded in 1965, compared with 7,696 in 1996. The impact of these trends on career opportunities for young PhDs is examined in detail in the chapter 3.

REFERENCES

Finn MG, Pennington LA, Anderson KH. 1996. Who stays–who leaves: foreign PhDs eight years after gaduation. Pesented at 1996 American Association for the Advancement of Science annual meeting.

NRC (National Research Council). 1978. A century of doctorates: data analysis of growth and change. Washington, DC: National Academy Press.

NRC (National Research Council). 1995. Study of research doctorate programs in the United States: Continuity and Change. Washington, DC: National Academy Press.

NSF (National Science Foundation). 1995. Survey of graduate students and postdoctorates in science and engineering. Washington, DC: NSF.

Van Ryzin G, Dietz S, Weiner J, Wright D. 1995. The employment outlook in the microbiological sciences, 1995. http://www.asmusa.org/pasrc/empoutlk.pdf.

3 EARLY-CAREER EMPLOYMENT PROFILES OF LIFE-SCIENCE PHDS

This chapter presents national survey data on the early-career employment of life-science PhDs over a 22-year period. The employment trends discussed here, combined with supply and demand indicators described in other chapters, constitute the basis of the committee's findings regarding the prospects for persons interested in pursuing careers in the life sciences. The survey data help to explain–and put in perspective–much anecdotal information that has come to committee members' attention about an apparent lack of employment opportunities for recent PhD recipients in the life sciences.

The figures in this chapter (and the tables in appendix F) document what fractions of life-science graduates held faculty, industry, and other positions within 10 years of earning their doctorates and how these fractions changed from 1973 to 1995. The committee presents the data with a focus on the fraction of PhDs holding each type of position rather than total numbers because fractions permit more precise comparison of opportunities available to students in various cohorts. Data on total numbers in different positions are presented at the end of the chapter and in the appendixes. The need for data on employment patterns was noted in a 1995 national study (COSEPUP 1995) that examined graduation education in all fields of science and engineering:

> Graduate scientists and engineers and their advisers should receive more up-to-date and accurate information to help them make informed decisions about professional careers; broad electronic access to such information should be provided through a concerted nationwide effort.

The importance of such information was also stressed by several speakers at a public meeting that the committee held in April 1996 and by many young scientists who have complained that they were unaware of the declining career prospects in their fields when they entered graduate school. Some of the latter felt that they had been misled by their mentors, who had conveyed an unrealistically optimistic view of the chances of obtaining faculty positions at major research universities. One explanation for the misinformation is that employment prospects in the life sciences have changed substantially over the last 2 decades; opportunities available to PhD recipients and postdoctoral scientists in the 1990s are different from when their mentors completed graduate training. The employment-progression matrices presented at the end of this chapter and the analyses that follow describe early-career profiles, which should be useful to faculty mentors and to the students and postdoctoral scientists whom they counsel.

Most of the data presented in this chapter come from the biennial Survey of Doctorate Recipients, which since 1973 has collected current employment information in a carefully selected sample (8-13%) of all PhD scientists and engineers in the workforce. Because the survey results are based on a relatively small longitudinal sample, reliable estimates are not available for narrow segments of this population. For example, one would like to be able to distinguish among patterns in different fields–construct separate career profiles of biochemists, plant biologists, epidemiologists, and so on. One might also like to examine the employment histories of minority group scientists and foreign students. Although the sample size does not permit such detailed analyses, it does provide comparisons of the career patterns of women and men and of the graduates of the 26 leading universities and other life-science PhDs.[1] In addition, an analysis has been made of the employment histories of graduates in biomedical and nonbiomedical fields. Because it is difficult to obtain reliable data on foreign nationals with temporary visas who receive their PhDs in the United States and say that they will remain in this country, the tables and figures presented in this chapter and appendix F include only US citizens and those holding permanent visas who had received life-science PhDs from US universities. Nevertheless, it should be recognized that a growing number of foreign students have taken postdoctoral appointments at US institutions and that many of them subsequently seek permanent employment here.

Despite the limitations described above, the analyses that follow provide valuable insights into how the employment opportunities have been changing over the last 2 decades. This historical picture is especially important in showing that the career options of today's students are different from the opportunities that their mentors had when they were in graduate school. This information has already proved useful to the committee in formulating its study findings and recommendations, but it might be of greater interest to graduate students, postdoctoral fellows, and faculty. The committee cautions, however, that the national picture of all life-science PhDs presented here does not necessarily apply to students in a particular field or university department. For example, only a small fraction of biostatistics graduates take postdoctoral appointments, whereas most biochemistry PhD recipients pursue postdoctoral training. Important differences might also be found among programs within the same field. The committee urges prospective students and postdoctoral fellows to seek detailed career information from the programs that they are considering and to compare this information with the national data presented in this chapter.

FACULTY POSITIONS

The most important change in the career patterns of life-science PhDs in the 22-year period was a steady decrease in the fraction holding tenure-track faculty positions. The decline, illustrated in figure 3.1, was observed in all PhD cohorts. For the youngest graduates (those 1-2 years after receipt of the PhD), the fraction holding faculty jobs fell from 0.4 in 1973 to 0.14 in 1995. Some of the precipitous drop might be explained by an increase in the fraction of graduates taking postdoctoral appointments during this period. However, a sharp decline was observed in the oldest cohort (9-10 years after PhD) as well. Only 39% of the latter group held faculty positions in 1995, compared with 61% 22 years earlier.[2] What might be most remarkable about this trend is the consistency with which it has occurred over the last 22 years.

It is important to recognize that a substantial decline in faculty opportunities was observed in PhD-granting universities, as well as in other academic institutions. In 1995, for example, only 34% of the graduates with 9-10 years of post-PhD experience held tenure-track faculty appointments in doctoral institutions; in 1973, the comparable figure was 47% (see table F.1). If this decline continues, fewer than one-third of the life-science students now completing their graduate training can expect to obtain tenure-track faculty positions in doctorate-granting institutions, which in the past have been the principal employers of PhDs in this field.

---

[1] In addition to the above limitations, a few caveats pertain. During the 1973-1995 survey period there have been some modifications in the sampling frame and the wording of specific questions asked in the survey. With regard to the former, the survey sample size was substantially reduced in 1991 (from about 13%-8%), and a concerted effort was made to improve the response rate, which rose from 55% in 1989 to more than 75% in later surveys. It is difficult to estimate the effect of this change on the survey results.

[2] It should be noted, however, that the total number holding faculty positions has substantially increased during the 22-year period (see figure 3.14).

Figure 3.1 Fraction of US life-science PhDs holding faculty positions, 1973-1995.

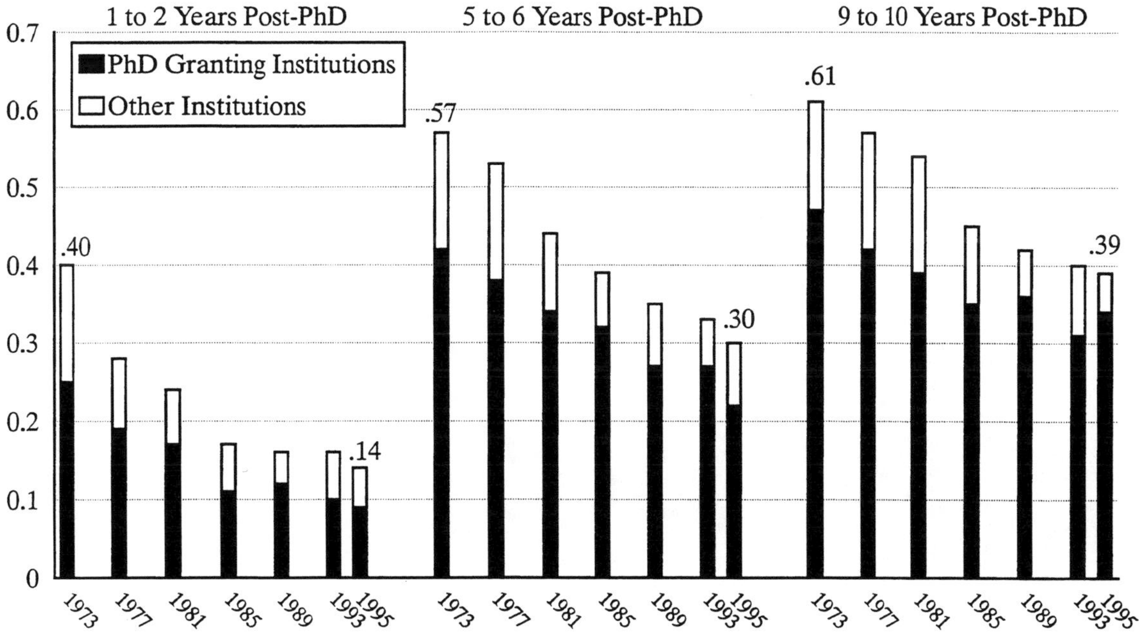

Data from table F.1.

EMPLOYMENT IN GOVERNMENT

A modest decline was also observed in the fraction of life-science PhD recipients employed in national laboratories and other federal, state, and local government positions. In 1995, only 11% of the 9-10-year cohort held government jobs, compared with 14% 22 years earlier (see figure 3.2). The decline might be attributed primarily to downsizing in the major federal laboratories, which in the past had employed large numbers of PhD scientists.

POSITIONS IN INDUSTRY

The appreciable decline in the fraction of young graduates taking faculty or government positions was accompanied by increased hiring in the industrial sector, especially among the more experienced graduates (see figure 3.2). In 1995, 23% of the life-science graduates with 9-10 years of experience were employed in industry, compared with only 12% in 1973. If that trend continues for the next decade, today's graduate students are more likely to find jobs in industry than on university faculties. However, it should be noted that most of the increase in industrial hiring occurred during the 1980s with only modest growth since 1989. Future employment opportunities in this sector will most likely depend on the national economy and in particular on the health of the biotechnology industry; both are difficult to predict with any confidence.

Figure 3.2 Fraction of US life-Science PhDs holding jobs in government, industry, and other sectors, 9-10 years after receipt of degree, 1973-1995.

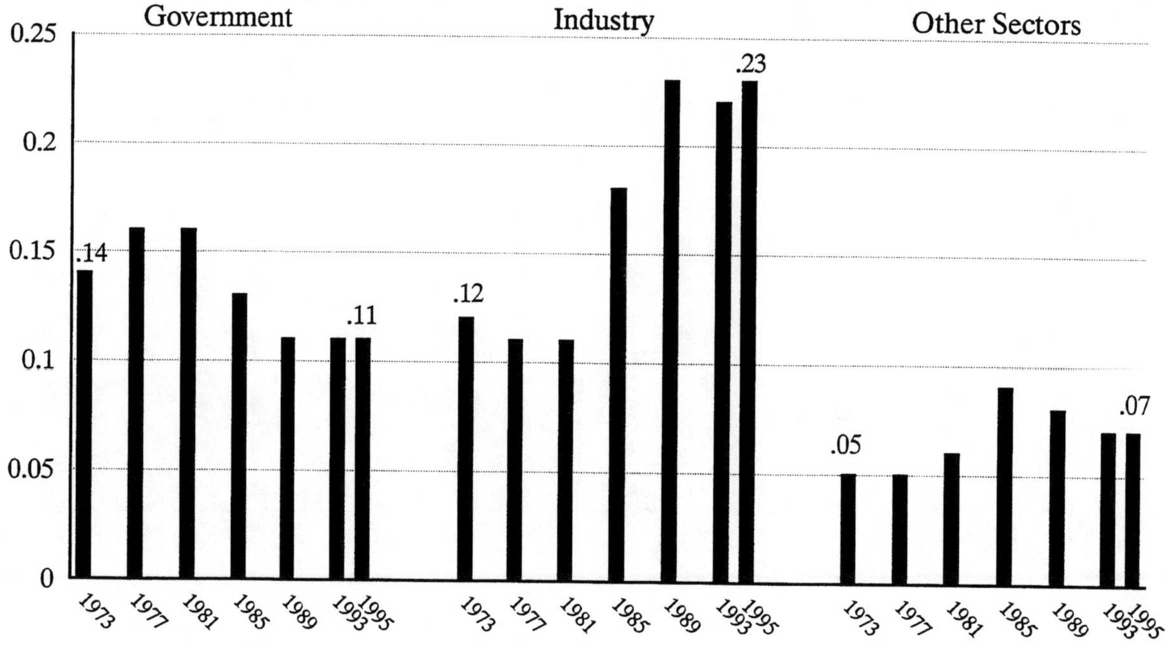

Data from table F.1.

OTHER EMPLOYMENT

Considerable attention has recently been given to "alternative careers" for PhD scientists (such as precollege teaching[3], and science journalism), but the fraction employed in such positions remained small. As shown in figure 3.2, only 7% of the life-science PhDs in 1995 held full-time positions outside academe, industry, and government, and—more important—the percentage has declined slightly over the last decade. Various alternative career opportunities (not involving research) might be available, but they are unlikely to be attractive to most young scientists who have just completed 10 years or more of predoctoral and postdoctoral training.

POSTDOCTORAL APPOINTMENTS

In addition to the growth in industrial employment, we observed a substantial increase in the number of graduates taking postdoctoral appointments in universities and in federal and industrial laboratories. As illustrated in figure 3.3, the fraction of life-science PhDs holding postdoctoral appointments 1-2 years after receipt of their doctorates more than doubled from 1973 to 1995, from 21% to 53%. Perhaps even more important is the increase in postdoctoral fellows in the older cohorts. In 1995, 29% of the graduates with 3-4 years of post-PhD experience

---

[3] For a discussion of the employment opportunities for PhDs in precollege teaching, see chapter 4 and COSEPUP 1995, p. 33-4.

Figure 3.3 Fraction of US life-science PhDs holding postdoctoral appointments in academe, government, and industry, 1973-1995.

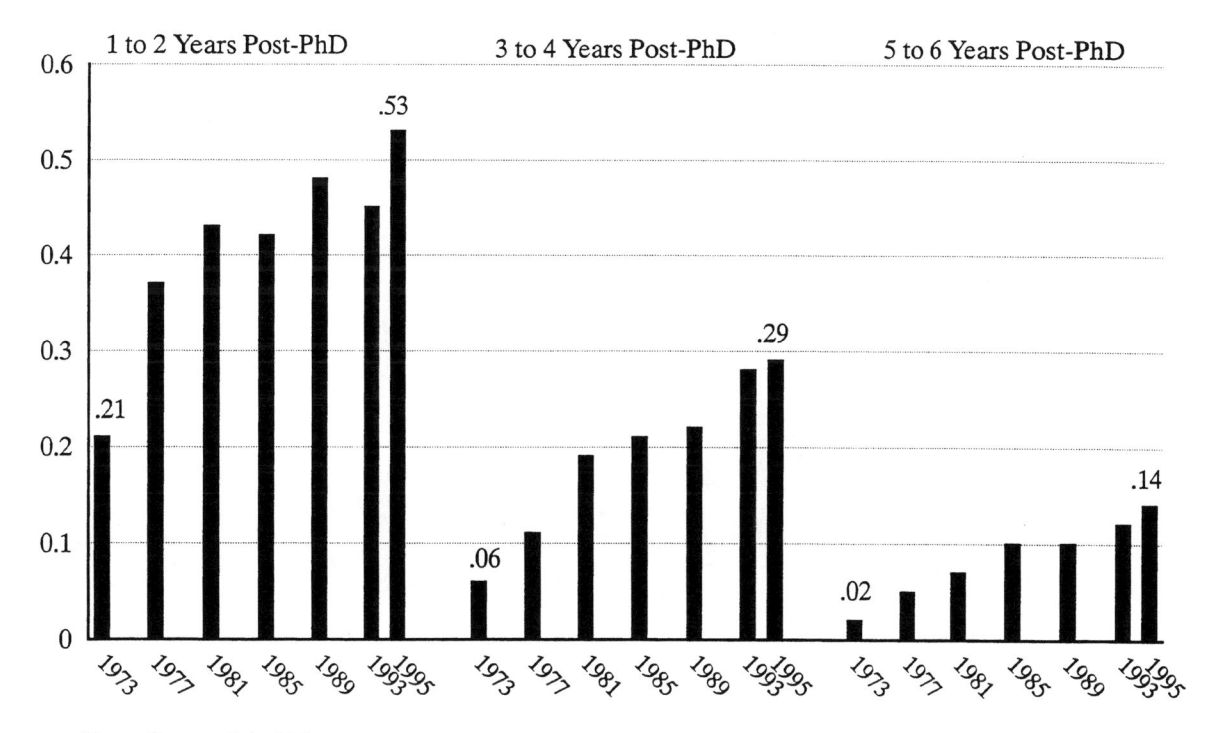

Data from table F.1.

and 14% of those with 5-6 years of post-PhD experience still held postdoctoral appointments, compared with only 6% and 2%, respectively, 22 years earlier. The availability of postdoctoral appointments has allowed young scientists to use their research training, even during periods when their immediate employment prospects were not very promising; that is, this apprenticeship has served as an "employment buffer". Nevertheless, the uncertainty (lack of job security) and low salary associated with these temporary positions might well explain the discontent and frustration that the committee has observed in young scientists who after 10 years or more of research training have not yet found permanent jobs. By "permanent" we mean positions in which young scientists can independently apply their education and training in positions that are not transitional, as postdoctoral fellowships, research assistantship, and associate positions generally are.

INVOLVEMENT IN RESEARCH

The fraction of young life-science PhDs who designated basic or applied research as their primary work activity grew substantially from 1973 to 1995. For the youngest cohort, the trend might be partly explained by the rapid rise in postdoctoral scientists, who devoted their full energies to research. However, even those with 9-10 years of post-PhD experience exhibited an increasing involvement in research–58% designating it as their primary activity in 1995 compared with 41% in 1973 (see figure 3.4). One may conclude from this finding that, despite a decline in the fraction employed on university faculties and in government, a growing majority of life-science PhDs have been fully using their research training–Although many might be postdoctoral fellows who are not independent researchers.

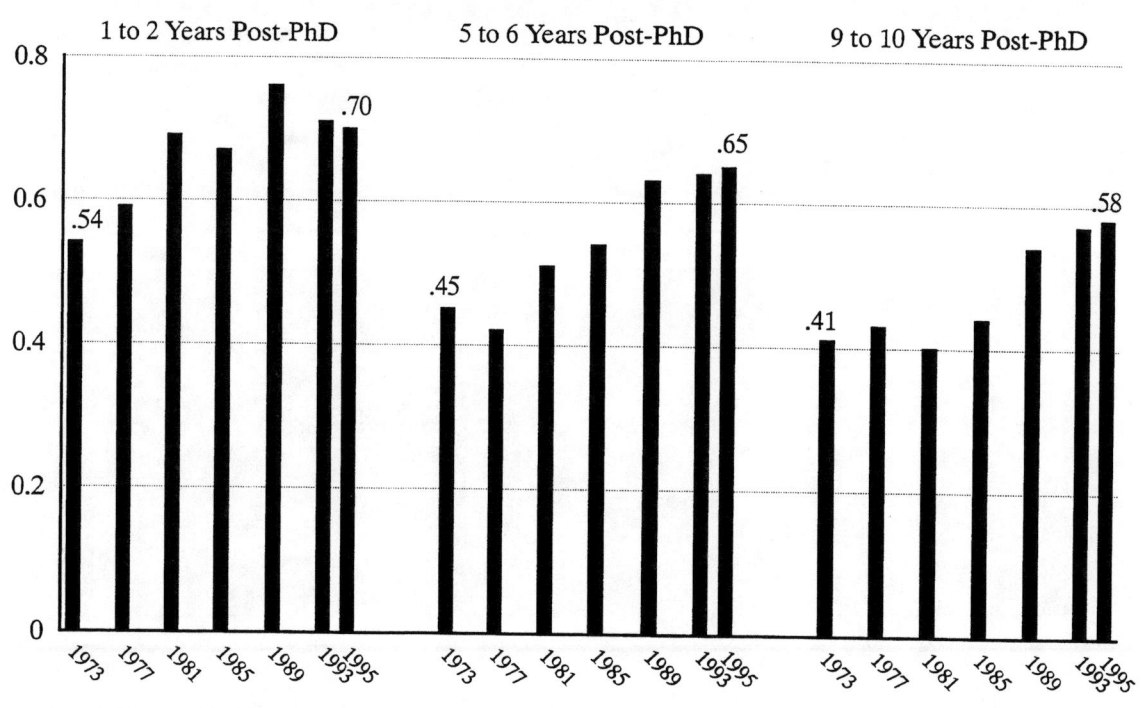

Figure 3.4 Fraction of US life-science PhDs involved primarily in basic or applied research, 1973-1995

1 to 2 Years Post-PhD | 5 to 6 Years Post-PhD | 9 to 10 Years Post-PhD

Data from table F.1.

UNEMPLOYMENT AND UNDER USE

Most life-science PhDs have been employed full-time in science and engineering endeavors. Data in figure 3.5 confirm that the unemployment rates for these young graduates averaged 1-2% during the 22-year period, and the fraction working part-time remained almost as low. Furthermore, no convincing evidence was found that an increasing fraction of young life-science PhDs are leaving the field.[4] The findings, when considered with the growing research involvement described above, suggest that employment prospects have been better for young PhDs in the life science than for graduates in many other sciences, such as mathematics, physics, and chemistry (COSEPUP 1995).

CAREER PATTERNS OF WOMEN AND MEN

Differences in the employment patterns of women and men narrowed during the 22-year period. As shown in figure 3.6, women with 9-10 years of post-PhD experience in 1973 were much less likely than their male colleagues to hold faculty appointments in doctorate-granting universities or to be employed in industry or government; but women were more likely than

---

[4] The 1993 and 1995 fractions working outside science and engineering fields, which are somewhat higher than in preceding years, are based on a new survey question and might not be comparable with earlier survey results.

Early-Career Employment Profiles of Life-Science PhDs

Figure 3.5 Fraction of US life-science PhDs unemployed, employed part-time, or employed outside science and engineering, 1973-1995.

Data from table F.1.

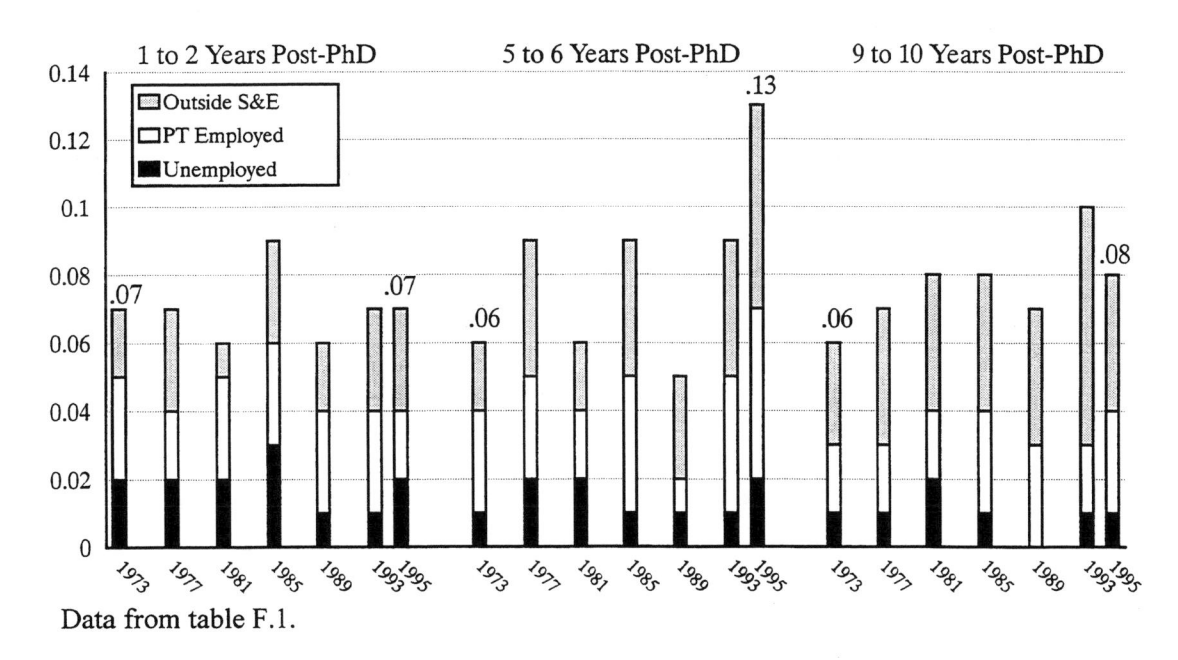

Figure 3.6 Fraction of female and male US life-science PhDs in faculty, industry, and government 9-10 years after receipt of degree, 1973, 1985, and 1995.

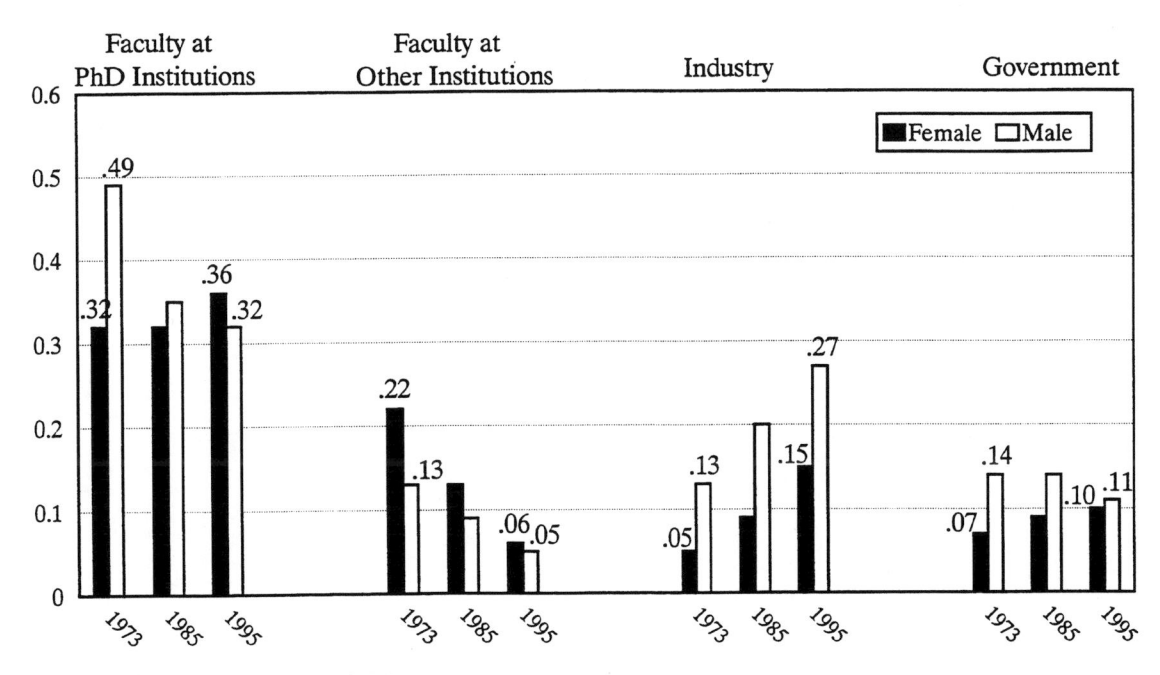

Data from tables F.2 and F.3.

men to hold faculty positions in 4-year and 2-year colleges. By 1995, however, most of those differences in employment situations had greatly diminished. Perhaps most striking is the finding that during the 22-year period the fraction of women with faculty appointments in PhD institutions actually increased slightly (from 32% to 36%) while the comparable fraction for men plummeted (from 49% to 32%). One important difference persisted: in 1995, men were nearly twice as likely as women to hold jobs in industry– 27% and 15%, respectively.

In 1973, women were much more likely than men to hold postdoctoral appointments (see figure 3.7). By 1995, the difference had greatly diminished. Nevertheless, it is important to recognize that the fraction involved in postdoctoral training increased substantially among both women and men during the 2 decades and that both sexes were spending, on the average, much longer periods as postdoctoral fellows. One large difference in employment status did not change much: women were still much more likely than men to be employed part-time. In 1995, for example, 7% of the women who had earned doctorates 3-4 years earlier worked part-time, compared with only 1% of the men.

Figure 3.7 Fraction of female and male US life-science PhDs holding postdoctoral appointments or part-time jobs 1-2 and 3-4 years after receipt of degree, 1973, 1985, and 1995.

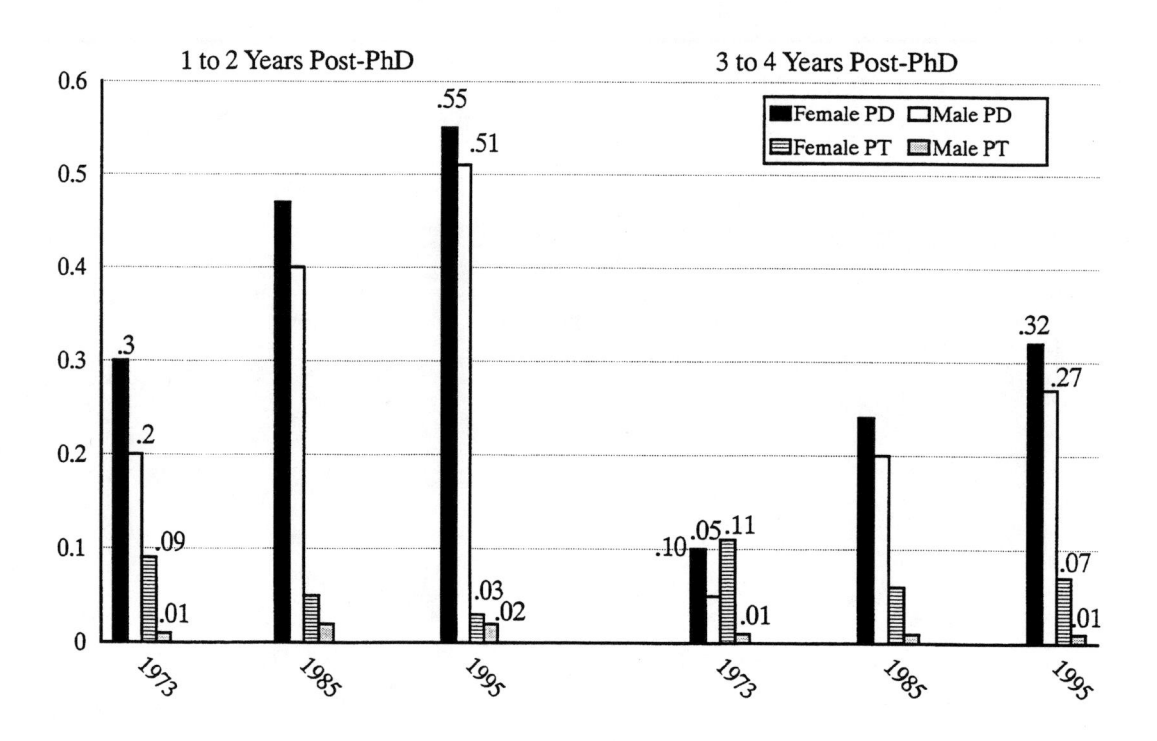

Data from tables F.2 and F.3.

GRADUATES OF TOP-RATED INSTITUTIONS

To compare the career patterns of life-science PhD recipients from the most prestigious programs with those from other schools, the survey sample was divided into two groups based on the reputational ratings (see footnote 2 in chapter 2) of the doctoral institutions. As shown in figure 3.8, graduates of the 26 top-rated schools were less likely than their colleagues–9-10 years after receipt of their PhDs–to hold positions in industry and government. What might be most important, however, are the 1973-1995 changes in the fraction with faculty appointments in doctorate-granting universities. The diminishing opportunities for such positions affected both groups of graduates, but those of the highest-rated institutions appear to have faired much better. In 1995, 45% of the latter graduates held faculty positions at PhD-granting universities, compared with 29% of the PhD recipients from other schools. In 1973, the differences between the too groups were negligible.

Figure 3.8 Fraction of US life-science PhDs from 26 highest-rated universities holding jobs in selected sectors, compared with PhDs from other institutions, 9-10 years after receipt of degree, 1973, 1985, and 1995.

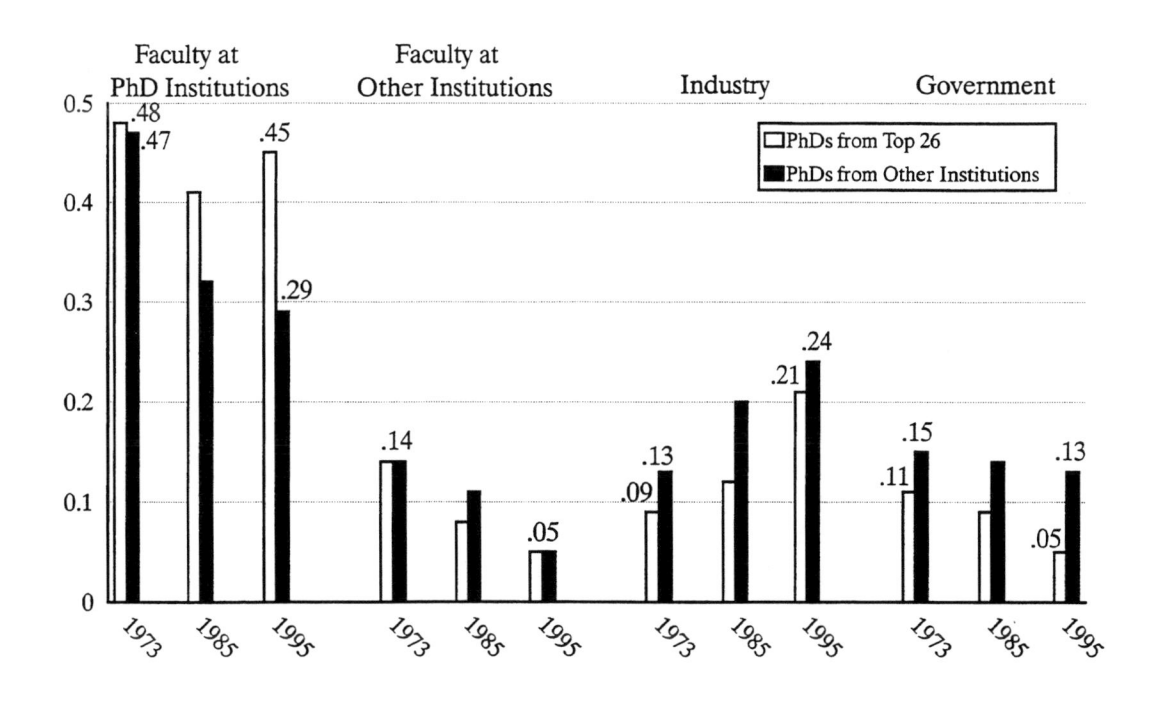

Data from tables F.4 and F.5.

Those from the highest-rated schools were also more likely to take postdoctoral appointments (see figure 3.9). In 1995, for example, 60% of the most recent graduates from the top-26 institutions held postdoctoral appointments, compared with 50% of the PhD recipients from other schools. That finding is not surprising inasmuch as graduates of the most prestigious programs were more likely than their colleagues to obtain university faculty positions, which usually require postdoctoral experience. Nevertheless, it should be noted that in 1995 17% of the PhDs from the top-26 schools still held postdoctoral appointments 5-6 years after graduation–an indication that many were having difficulty in finding permanent positions.

FIELD DIFFERENCES

As already indicated, the size of the survey sample did not permit an analysis of employment patterns in individual disciplines. However, it was possible to divide the survey responses into two broad categories of fields: biomedical and nonbiomedical, as listed in appendix D. Although the general trends in employment were similar, the employment profiles of the two groups reveal some important differences (see figure 3.10). Biomedical PhDs were somewhat more likely than their nonbiomedical counterparts to hold faculty positions at PhD-granting institutions; those in the nonbiomedical fields were somewhat

Figure 3.9 Fraction of US life-science PhDs from 26 highest-rated universities holding postdoctoral appointments, compared with PhDs from other institutions, 1973, 1985, and 1995.

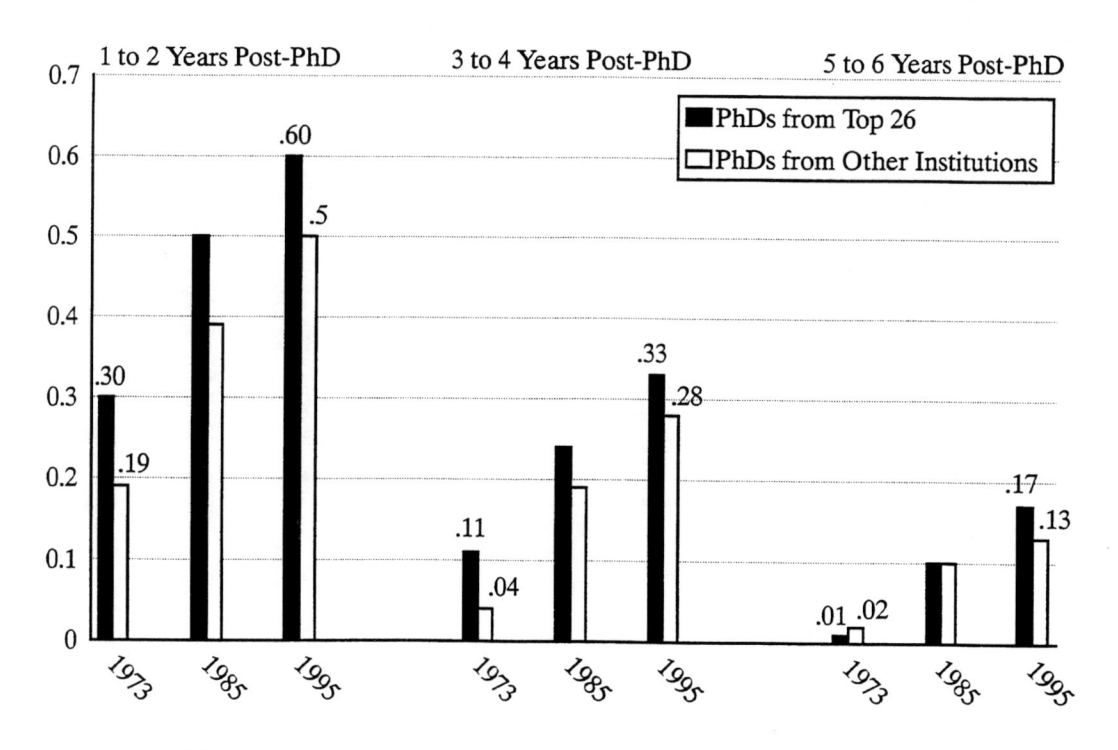

Data from tables F.4 and F.5.

Figure 3.10 **Fraction of biomedical and nonbiomedical US life-science PhDs in faculty, industry, and government 9-10 years after receipt of degree, 1973, 1985, and 1995.**

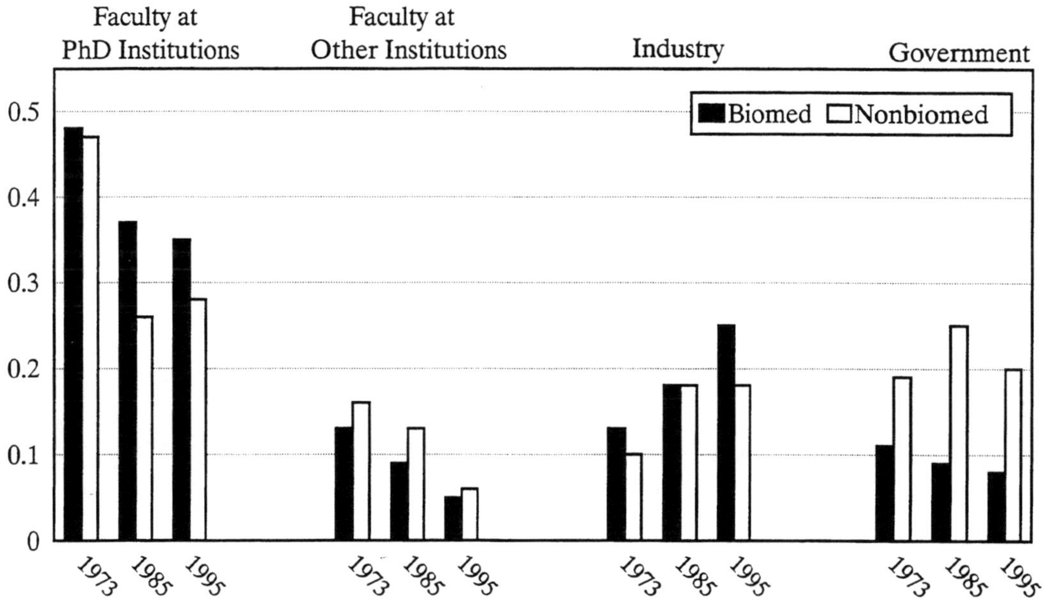

Data from tables F.6 and F.7.

more likely to hold faculty positions at other than PhD-granting institutions, Nonbiomedical PhDs were far more likely to find work in government than biomedical PhDs. There are temporal differences as well. For example, the fraction of nonbiomedical PhDs on the faculty of PhD-granting institutions increased slightly between the 1985 and 1995 surveys, whereas the fraction of biomedical PhDs in such positions continued the steady decrease begun in 1975. However, the number of nonbiomedical PhDs in the sample was only one-fifth the number of biomedical PhDs, and the differences might be more apparent than real. A high percentage of biomedical PhDs took postdoctoral positions in every year examined in this report. However, graduates in the nonbiomedical group increasingly also took postdoctoral positions: in 1995, 33% of those with 1-2 years of post-PhD experience held postdoctoral fellowships, compared with only 6% in 1973 (see figure 3.11). It appears that the trend toward more frequent and longer postdoctoral appointments affected all graduates–not just those in the biomedical sciences.

DISCUSSION

The foregoing analysis helps to interpret an important paradox that the committee has encountered. Young graduates in the life sciences have expressed frustration and anguish over the dearth of career opportunities available to them–especially in the academic sector, where often more than 100 candidates have applied for a single faculty opening–but there is no evidence of appreciable unemployment or underemployment.

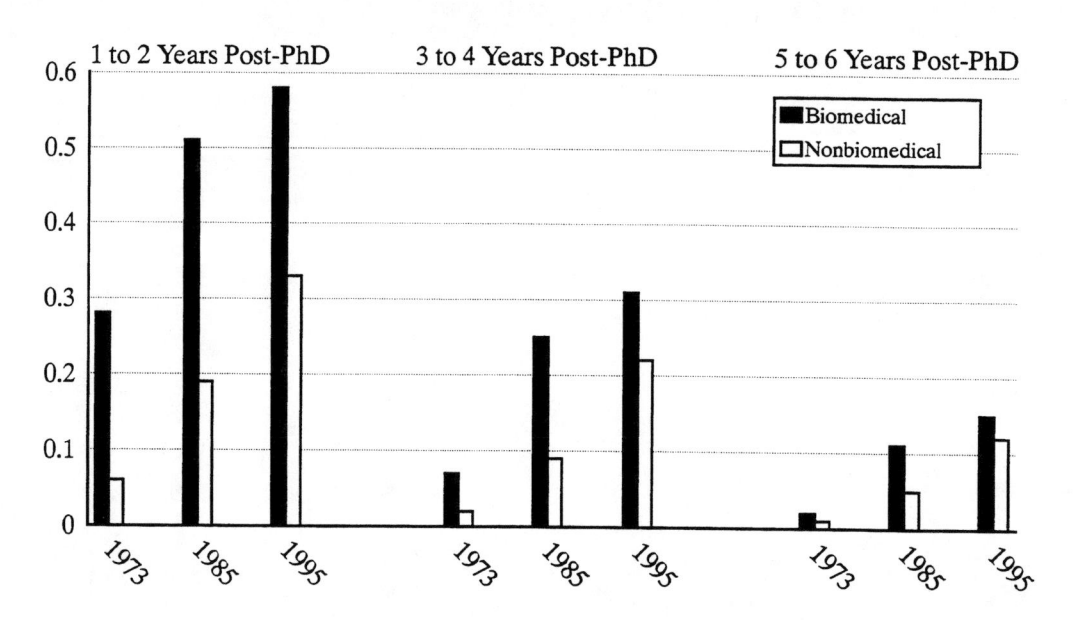

Figure 3.11   Fraction of biomedical and nonbiomedical US life-science PhDs holding postdoctoral appointments, 1973, 1985, and 1995.

Data from tables F.6 and F.7.

The data presented in this chapter confirm that the unemployment rate among recent PhD recipients in the life sciences has remained low (between 1% and 2%), and there is no indication that large numbers of them have left the field. Moreover, a majority of the graduates have been primarily engaged in basic and applied research–an indication that they have been fully using their research training–and this fraction has been rising. The intensive research involvement might be at least partly attributed to an expansion in industrial hiring, which began in the early 1980s, as well as to a large increase in the number of postdoctoral fellows.

So what is the problem? Over the last 2 decades, there has been a substantial decline in the fraction of young PhDs in the life sciences who have obtained tenure-track positions on university and college faculties–the positions considered most desirable by many life scientists. If the decline continues at its current rate, fewer than one-third of today's graduates can be expected to obtain faculty appointments, to which a majority of students have aspired. The apparent mismatch between career expectations and opportunities for faculty positions might be ameliorated, at least in part, by a growing awareness among students, postdoctoral fellows, and faculty of the career options available to today's graduates. It is the committee's hope that the career-progression matrices and accompanying analysis presented here will enhance their awareness of the changing employment prospects in the life sciences.

A second problem, perhaps more difficult to solve, is the increase in the fraction of young scientists who, after extensive postdoctoral apprenticeships, still have not obtained "permanent" full-time positions in academe, industry, government, or private research organization. As illustrated in figure 3.12, in 1995, 39% of life-science PhDs 5-6 years after receipt of their doctorates held postdoctoral fellowships or other nonfaculty jobs in universities, were employed

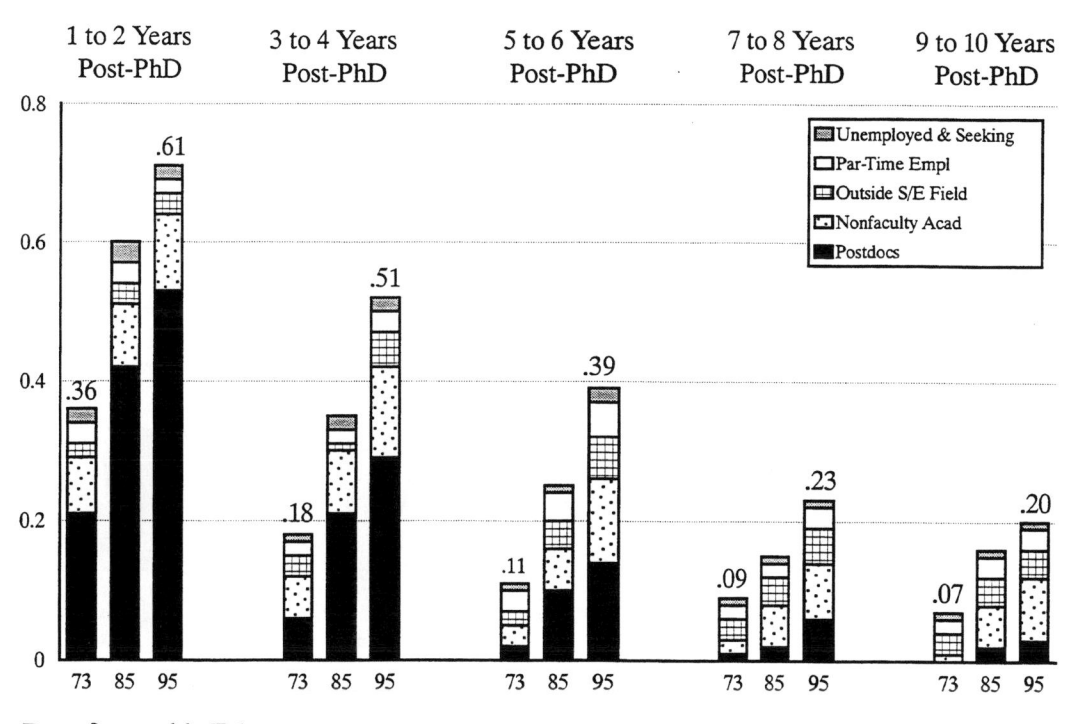

Figure 3.12 Fraction of US life-science PhDs not holding permanent full-time jobs in science or engineering, 1973, 1985, and 1995.

Data from table F.1.

part-time, worked outside science and engineering, or were unemployed; the comparable fraction in 1973 was only 11%.[5] What might be most alarming about the 1995 figure is that it reflects the situations of those who earned PhDs in 1989 and 1990. For those receiving their doctorates now, the prospects for finding career positions on university faculties or in government or industry where their long research training will be fully used are even less certain. For young scientists caught in this "postdoctoral holding pattern", the frustrations are understandable; most of them are 35-40 years old, and they typically receive low salaries and have little job security or status within the university setting (for example, most are not permitted to apply for research grants as independent investigators). Moreover, they are competing with a rapidly growing pool of highly talented young scientists–including many highly qualified foreign postdoctoral appointees–for a small number of jobs in academe, government, and industry. This situation–and its implications for individual scientists and the research enterprise–is a matter of great concern to the committee. We explore these implications in later chapters.

Although the prospects for permanent research positions have declined substantially for all life-science graduates, different groups have been unequally affected by the trend. As shown in figure 3.13, the fraction of women with 5-6 years of post-PhD experience who still held

---

[5] During the 22-year period, the total *number* in these types of positions quintupled.

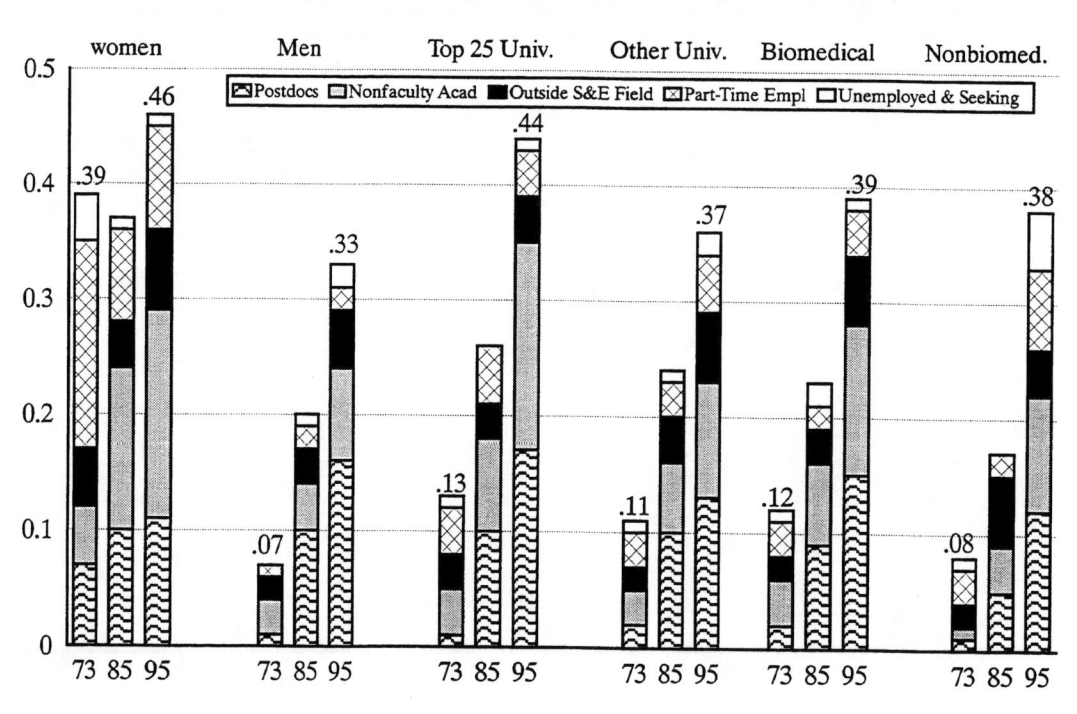

Figure 3.13 Fraction of 5-6 year cohorts not holding "permanent" full-time jobs in science or engineering, 1973, 1985, and 1995.

Data from tables F.3 and F.7.

"temporary" or part-time positions has been much higher than the fraction of men, but the difference narrowed from 1973 to 1995. Graduates of the highest-rated institutions found positions later than their colleagues from other schools. However, the difference might be explained primarily by the fact that graduates of the leading institutions were more likely to take postdoctoral apprenticeships and more likely to hold postdoctoral or other nonfaculty positions in academe 5-6 years after graduation. Similarly, recent biomedical-science PhDs were more likely than graduates in nonbiomedical life-science disciplines to hold temporary (non-tenure-track) appointments in universities. Those and other differences in the career patterns of individual groups indicate that the observed national employment patterns of all life-science PhDs do not necessarily apply to those in a particular field, department, or group. For that reason, it is imperative that the employment histories of graduates of individual university departments be made available to prospective graduate students and postdoctorals.

The changes in career prospects for young scientists occurred while the total numbers of life scientists in the workforce continued to increase. Figure 3.14 shows the numbers of life-science PhDs (US citizens and permanent residents only) in the workforce. The figure reveals that the numbers employed in every sector continued to grow throughout the 22-year period. Much of the growth in the faculty at PhD-granting institutions occurred before 1989. In contrast, the most pronounced and persistent trend in the 22-year period is the growth in the numbers in industry, post-

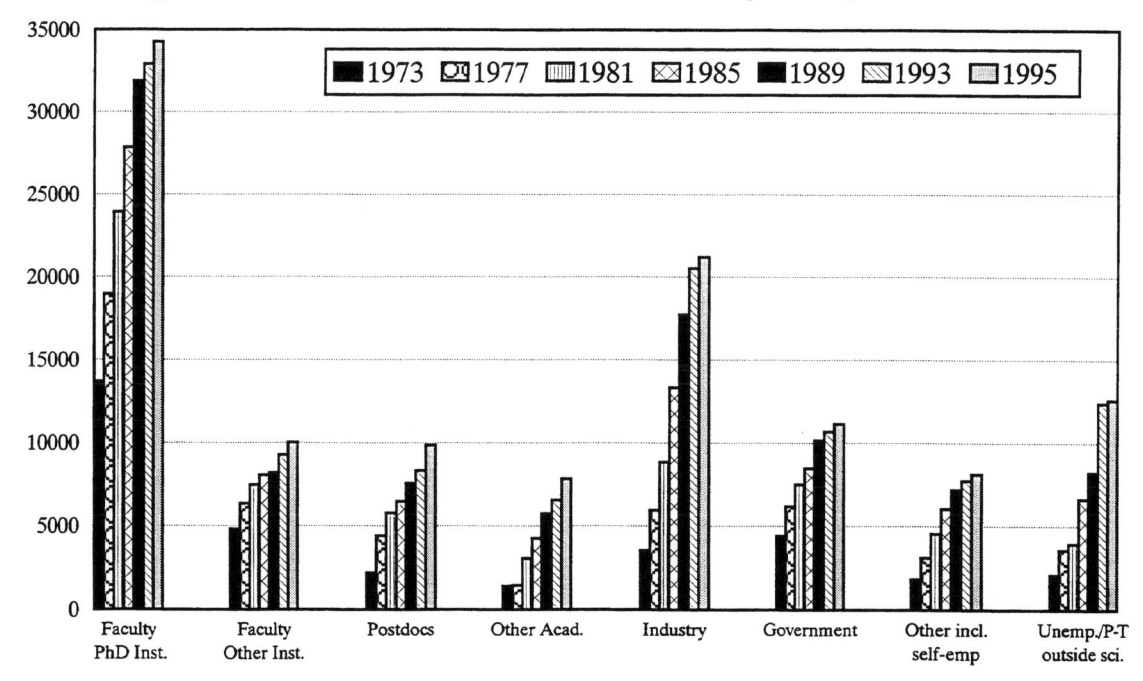

Early-Career Employment Profiles of Life-Science PhDs

Figure 3.14 Number of US life-science PhDs by sector, 1973-1995.

Data from table F.8.

doctoral fellows,[6] other (nontenured or non-tenure-track) academics, other including self-employed, and the group containing unemployed, part-time, and PhDs now working outside science and engineering.

The results of the changing employment patterns are illustrated in figure 3.15. The figure shows changes in the number of life scientists employed in each sector–or unemployed and seeking employment–in three periods: 1973-1981, 1981-1989, and 1993-1995, the latest period on which data are available. In the 1970s, by far the largest increase in the workforce was in faculty jobs (41.5% of the total growth); in the 1980s, industrial positions accounted for the largest share of additions to the workforce (28.1% of the total growth), just ahead of faculty positions. However, in 1993-1995, the total growth in faculty and industry workforce was less than the increase in the numbers of persons in temporary and under use positions (postdoctoral and other nonfaculty staff, unemployed, part-time employed, and outside science and engineering), which accounted for 45.4% of the growth in life-science "workforce", compared with about 25% in earlier years. The data in figure 3.15 help to explain the conundrum of a growing workforce, a low rate of unemployment, and a high level of dissatisfaction among life scientists seeking to establish careers. Compared to previous years, an increasing percentage of these younger life scientists are in temporary positions.

---

[6] Figure 3.14 used Survey of Doctorate Recipients data, which include only US citizens and permanent residents. The numbers of postdoctoral fellows shown in the figure are therefore lower than the numbers shown in chapter 2.

Figure 3.15 Increase in life-science PhD workforce in 1973-1981, 1981-1989, and 1993-1995, by sector.

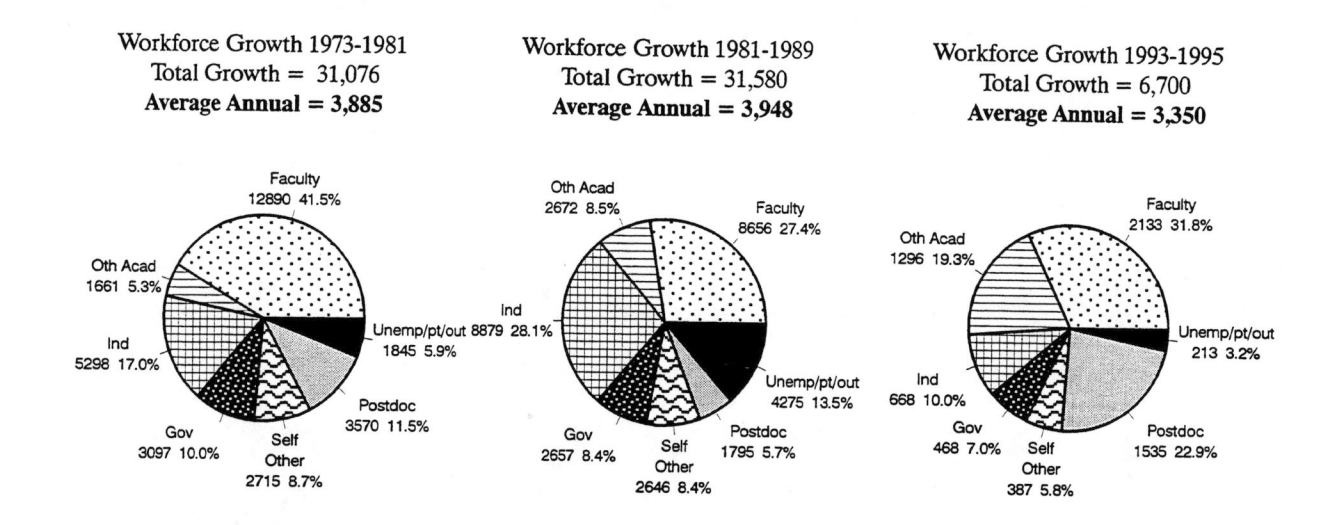

Workforce Growth 1973-1981
Total Growth = 31,076
Average Annual = 3,885

- Faculty: 12890, 41.5%
- Oth Acad: 1661, 5.3%
- Ind: 5298, 17.0%
- Gov: 3097, 10.0%
- Self Other: 2715, 8.7%
- Postdoc: 3570, 11.5%
- Unemp/pt/out: 1845, 5.9%

Workforce Growth 1981-1989
Total Growth = 31,580
Average Annual = 3,948

- Oth Acad: 2672, 8.5%
- Faculty: 8656, 27.4%
- Ind: 8879, 28.1%
- Gov: 2657, 8.4%
- Self Other: 2646, 8.4%
- Postdoc: 1795, 5.7%
- Unemp/pt/out: 4275, 13.5%

Workforce Growth 1993-1995
Total Growth = 6,700
Average Annual = 3,350

- Faculty: 2133, 31.8%
- Oth Acad: 1296, 19.3%
- Ind: 668, 10.0%
- Gov: 468, 7.0%
- Self Other: 387, 5.8%
- Postdoc: 1535, 22.9%
- Unemp/pt/out: 213, 3.2%

Data from table F.8.

REFERENCES

COSEPUP (National Academy of Sciences, Committee on Science, Engineering, and Public Policy). 1995. Reshaping the graduate education of scientists and engineers. Washington, DC: National Academy Press.

4 OPPORTUNITIES, CONSTRAINTS, AND FUTURE NEEDS

The future promises many exciting opportunities for scientific research in the life sciences, but there are also considerable uncertainties. This chapter briefly identifies some of the newly emerging fields of the life sciences that hold particular promise for the immediate future. It then describes some of the uncertainties that life scientists will face and concludes with a discussion of the diversity of career options that might be available to young life scientists now and in the future.

EXCITING EMERGING FIELDS OF INQUIRY IN THE LIFE SCIENCES

Research in the life sciences is high on our nation's list of priorities largely because of the likelihood that this research will improve the well-being of our population. Of the many promising fields of science that will contribute to economic and social well-being, we mention here only a few examples.

NEUROSCIENCE

The 1990s have been called the "decade of the brain", and neuroscience offers essentially unlimited challenges and opportunities in both basic and applied research. High on the list of promising fields of research is the quest for links between cognition and the molecular activity of memory processes in the brain. New concepts and new techniques are opening exciting research opportunities. For example, neuroscientists are using state-of-the-art genetic engineering, imaging methods, and monitoring of brain-cell physiology to define the molecular bases of memory, recognition, and learning in experimental animals. The molecular mapping and elucidation of complex brain-cell functions will advance the understanding of Alzheimer's disease, learning disorders, addiction, and other medical and psychological conundrums that currently plague society. Careers in the neurosciences can be based on training in many combinations of molecular biology, neurobiology, physiology, psychology, and computer science.

GENE THERAPY

Gene therapy is based on the transfer of genetic material into a human. Gene delivery can be accomplished either directly by the administration of gene-containing viruses or DNA to blood or tissues or indirectly through the introduction of cells that have been manipulated in the laboratory to harbor foreign DNA for the purpose of treating disease. By altering the genetic material of somatic cells, gene therapy could correct underlying disease-specific pathophysiologic characteristics. In some instances, it offers the potential of a one-time cure for devastating, inherited disorders, such as diabetes. In principle, gene therapy should be applicable to many diseases for which current therapeutic approaches are ineffective or when the prospects for effective treatment appear exceedingly low. As of June 1995, 106 clinical protocols involving gene transfer had been approved by the National Institutes of Health (NIH) Recombinant Advisory Committee (RAC). Indeed, more than 600 human subjects have already undergone gene transfer experiments. NIH provides about $200 million per year for research related to gene therapy, and industrial support of gene-therapy research has grown steadily. Industry now exceeds NIH in funding and underwrites most of the approved clinical protocols. This young field is a frontier of modern medicine, open to people with MD or PhD degree in molecular genetics, molecular biology, or related sciences.

STRUCTURAL BIOLOGY

All of genetic information in an organism is encoded in the DNA or RNA sequence of its genome. The genome projects that are now under way are producing vast amounts of data that will be essential for understanding the normal and pathologic physiology of humans and of the many plants and animals on which our lives depend. There are, however, many unsolved problems related to genome research, some of which are so novel that they are only now being defined as specific subjects for research. For example, how is gene expression regulated on the molecular level? How does chromosomal architecture influence the rate of gene expression? How is the three-dimensional structure of proteins defined by the amino acid sequences that are specified by the genome? What are the mechanisms of protein-protein recognition in complex biochemical processes? What processes regulate the assembly of protein complexes into organelles?

Structural biology provides some of the research tools that are necessary to solve those grand challenges in molecular and cellular biology. Current research is providing improved techniques by which to determine the high-resolution structures of macromolecules, and these methods are being used to study processes of molecular recognition, signal transduction, allosteric regulation, and protein folding. The resulting data are often of immediate practical value for such undertakings as rational drug design. They are also of fundamental theoretical value as thermodynamic and kinetic data become available to complement the structural information. The resulting synergy between different kinds of molecular data is providing the views that will be necessary to understand complex biologic processes. This critically important line of inquiry is now in its earliest stages, and considerable effort will be required to realize the practical benefits of such research. A person interested in a career in structural biology should obtain a PhD degree in biochemistry, biophysics, or structural and computational biology. Prerequisites include a strong background in computer science and physics, chemistry, biology, or mathematics.

BIOINFORMATICS

Bioinformatics uses computer technology to solve informational problems in the life sciences, for example, the identification of DNA sequences in the human genome that are markedly similar to genes that have been identified and studied in experimental organisms such as yeasts. The computer databases of genome and protein sequences are now large enough to require new models for the analysis and comparison of biologic systems, and new algorithms are under development to integrate heterogeneous data into coherent programs. Informatics also plays a role in modeling the interactions between drugs and proteins or physiologic processes, in the diagnosis of disease, and in keeping track of huge databases, from the DNA sequences cited above to records of patient care.

Medicine is an information-based art and science, and the opportunities for computer applications are constantly expanding. Three-dimensional visualization of human anatomy is already an instructional tool, and the visual modeling of changes in tissue structure during disease progression offers parallel opportunities. Large pharmaceutical houses are especially interested in scientists with training in bioinformatics, given the explosion of new data from large-scale sequencing projects, like the work on the human genome, which will require new technologies for information processing to assist in the exploitation of data for product development. Young people with advanced training in statistics, information theory, artificial intelligence, and other aspects of computer science can make major contributions.

ENVIRONMENTAL ISSUES

The growth of human populations is an important driving force in the accelerating changes that are occurring in the managed ecosystems on which we depend for food, fiber, and services, such as the maintenance of clean air and water. Human activities are measurably changing the composition of the atmosphere, adding carbon dioxide and methane, which alter the radiative balance of the planet, and chlorine gas, which destroys the ozone layers in the stratosphere. Humans have already destroyed vast tracts of tropical forests and agriculturally productive land. Industrial and human wastes have degraded some of the largest sources of fresh water. We are witnessing the rapid extinction of many species and the introduction of pests and infectious organisms into new environments, sometimes with calamitous results. There is an obvious need for increased attention to these problems and for research to find their solutions. Scholars who are expert in all aspects of environmental sciences will be required to understand the increasing stresses placed on the environment by the expanding human population and the concomitant growth of industry. Careers in this challenging field will require training in population biology, ecology, the social sciences, and related agriculture sciences.

BIOLOGIC CONTROL OF PLANT PESTS

The major increases in agricultural productivity that followed World War II were attributable in part to the widespread use of synthetic chemical pesticides for the control of insects, weeds, and plant pathogens. Initial successes have been followed by unexpected consequences, including injurious effects on nontarget organisms, contamination of soil and water with chemical residues, and the development of pesticide resistance, particularly among insects. In addition, the potential harmful effects of pesticides in the food chain offer considerable reason for concern.

There is a growing consensus that pest-management systems based on biologic control agents will provide a more desirable approach for resolving some of the current problems and reducing the use of synthetic pesticides. Achieving a shift to biologic control agents will, however, require the development of treatment strategies that are inexpensive, are easily applied, offer little or no hazard for nontarget organisms (including people), are equal in efficacy to or better than current pesticides, and are predictable under a range of environmental conditions. The successes in developing biologic control systems for insects have not been matched in progress toward commercial biologic control of plant pathogens or weeds. Unfortunately, the knowledge that is necessary to develop such biologic control agents will require a massive expansion of current research effort, and it will involve the complete spectrum of basic and applied life sciences.

Many of the major corporations involved in development of disease-control agents have closed research laboratories that have a primary assignment in biologic control agents. Emphasis has shifted to transgenic plants with insect-control characteristics or chemicals that turn on resistance mechanisms when applied to plants. Extensive growth in this type of research is foreseen. Some of the plant diseases that are most recalcitrant to all known control efforts are caused by soilborne pathogens. A deeper understanding of the complexities of the physical and biologic components of soil will require research on the microflora and microfauna of the leaf and root systems of plants going well beyond the bounds of our current knowledge. Furthermore, biologic control agents that are highly effective under greenhouse conditions are often ineffective or unpredictable when tested in the field and in different geographic regions. Thus, it is likely that extensive field testing and modification will be needed to develop and market effective biologic products. This phase of development will require many more agricultural biologists than are available today.

AQUACULTURE

A different opportunity for expanded employment of life scientists will be found in aquaculture. There has been a dramatic decline in the productivity of fisheries around the world, and successful expansion of aquaculture will depend on increased knowledge about the diseases of fish, the application of improved breeding and selection procedures, and the nutritional requirements of fish under the controlled conditions of aquaculture systems. This is a comparatively unexplored field of modern biology in which much remains to be done.

PROSPECTS FOR RESEARCH FUNDING

It is difficult to predict how research funding will fare in the future. Just 2 years ago, in a mood of concern about reduction of the federal budget deficit, it was predicted that the budgets of federal research agencies might fall by up to 20%. In President Clinton's proposed budget for FY 1999, the planned increase for NIH is 8.4 and the increases proposed for the National Science Foundation and the Department of Energy are even higher. It is important to note that research budgets were not static from the late 1980s to the present. NIH regularly increased its budget by about 5% per year. But chapters 2 and 3 show that the large increase in the number of life-science PhDs resulted in decreases in the fractions of the PhDs who obtained "permanent" positions in academe, industry, and government research. Whether the increases proposed for FY 1999 will come about and whether increased funding will change the trends that we have reported is problematic. The mood in Washington continues to favor containment of discretionary expenditures.

On the national level, the shifting of responsibility for welfare expenditures to the states and the states' preoccupation with healthcare costs, prison costs, and their own financial situations, imply that state support for research is not likely to expand. Indeed, state support for public higher education has moderated under all those trends, and public higher education has increasingly been financed by tuition income rather than tax revenue.

Nongovernment sources of support clearly are important for basic life-science research and funds from private foundations, such as the Howard Hughes Medical Institute and the American Cancer Society or American Heart Association, will probably continue at the same or slightly increased levels. But private philanthropy does not have the resources to compensate for a substantial decrease in federal funding (Ruzek and others 1996). Alhough industry now spends more on life-science research and development than does the federal government, industrial research is targeted mostly at problems that are expected to yield commercial payoffs in the short run. Only the government is currently willing to take the long-range view that recognizes the tremendous returns offered over the years by investments in basic research. The basic life-science research enterprise must therefore assume that major increases in its grant support are unlikely.

CHANGES FACING HIGHER EDUCATION

The nation's research universities face increasing financial pressures that are forcing changes in priorities and shifts of resources to different academic purposes. Of special interest for this report is the impact of such reorganization upon university-based research in the life sciences. For the last 10-15 years, university operating costs have been rising rapidly—more rapidly in most instances than inflation (Clotfelter 1996). Every cost, from janitorial supplies to faculty salaries, has increased while increases in income have not kept pace. Below are some specific examples.

CHANGES IN THE FINANCING OF UNDERGRADUATE EDUCATION

Like all institutions of higher learning, research universities have accepted the responsibility of providing financial aid to undergraduate students from minority and disadvantaged populations. Many private universities have maintained policies of need-blind admission and need-based financial aid by drastically increasing the fraction of their resources that is devoted to this purpose. Except for the few universities that have very large per-student endowments, the funds for financial aid have come mainly from increases in tuition. Reliable studies estimate that 15-40% of tuition revenue is used for undergraduate financial aid at various private institutions. The steep increase in tuition has, however, begun to arouse public concern, if not resistance, and has put pressure on universities to limit future increases. Tuition at public universities too have been rising faster than inflation, as the share of educational costs supported by state governments has declined.

Increased attention to undergraduate education at research universities has resulted not only from these financial factors, but also from evidence that their clientele is becoming aware that some portion of undergraduate tuition has implicitly subsidized research. The intellectual justification for this subsidy is that undergraduate access to leading researchers is a unique feature of research universities. It follows that providing an attractive environment for research-oriented professors is a legitimate part of the cost of undergraduate education. The question remains open whether families will continue to accept this rationale for high tuition costs. Given the widespread resistance to further tuition increases and the competition between the legitimate goals of tuition remission and research, it is unlikely that substantial additional resources for basic work in the life sciences will come from the research universities themselves.

DIFFICULTIES IN RECOVERING THE COSTS OF EXTERNALLY SUPPORTED RESEARCH

At a typical private research university, only about 85% of the indirect costs of sponsored research has been recovered in recent years. The situation in public research universities is probably no different. The shortfalls result from the fact that many government agencies, as well as many private foundations and corporations, have refused as a matter of policy to pay full indirect costs for research. Other agencies, which negotiate indirect costs according to some formula, have required "cost-sharing" by the university; have refused to accept outside-the-formula "special studies", which justify above-average costs; or have placed non-negotiable "caps" on particular items in the indirect-cost pool, generally for the explicit purpose of limiting outlays for research grants. As budget-balancing continues to occupy center stage in Congress, research universities face a likely decline in their real levels of federal support.

To maintain an adequate volume of research and the infrastructure to support it (the object of indirect-cost recovery), research universities must find alternative sources for research funding. Although increased gift income is one possibility, undergraduate financial aid and research will probably continue to compete with one another for scarce tuition dollars, at least at private research universities. Successful efforts to maintain levels of research support will probably lead to fewer low-income students at these institutions. Alternatively, maintaining current levels of financial aid and student diversity will mean less internal support for research. Only if universities can achieve substantial cuts in other areas of costs can this tradeoff be avoided.

CHANGES IN RETENTION AND HIRING OF FACULTY

One of the principal components of a university's budget is faculty salary, there is a natural administrative interest in opportunities for savings in this line. Unfortunately for this purpose, the abolition of mandatory retirement at a designated age has narrowed one such opportunity: it appears that a substantial number of professors are choosing to retire at later ages. Even a modest increase of 3-5 years in age of retirement (to 68 or 70, instead of 65) will mean an increase of 10-15% in the mean duration of a faculty career and an equivalent decrease in the number of people who can enter that career, all other things being equal. That not only slows the rate of faculty replacement, but it increases salary costs because senior faculty tend to be more expensive than their younger colleagues. It is not yet clear what strategies might help to reverse this trend. Attempts by universities to do so, by offering incentives to retire, do not appear to have saved money in the short run.

The current faculty age distributions at almost all colleges and universities virtually guarantee that the coming years will see vacancies that can be filled by younger scientists. The situation does not, however, guarantee that there will be vacancies for research-oriented faculty, nor that the positions available will be tenure-track. Universities seem to be responding to financial pressures by hiring more nontenure and part-time faculty. The reduction in tenure-track opportunities might make academic research posts less attractive to young scientists and have an impact on the extent to which talented college students are drawn into life-science research.

CHANGES IN ACADEMIC HEALTH CENTERS

Medical schools, which are generally parts of research universities, now face additional problems in maintaining a healthy research environment. Academic health centers (AHCs) include basic-research faculty and clinical researchers, as well as medical educators and physicians; these scientists work collectively to provide teaching, research, and clinical care. AHCs emerged during the period of unprecedented growth in the health-care sector that followed World War II. Substantial resources became available for building health-care partnerships among medical schools, university hospitals, and private medical centers. The resulting AHCs deliver multiple health-care services.

AHCs have flourished on federal dollars, along with a steady stream of income from faculty practice plans. Indeed, some AHCs today receive over 50% of their income from revenues for patient care. Faculty practice plans in 1993 provided at least $2.4 billion in support of academic programs, including undergraduate medical education ($702 million), graduate medical education ($594 million), and other academic support ($244 million) (Jones and Sanderson 1996). Faculty research grants also provide income to AHCs in the form of faculty and staff salary support and indirect-cost recovery. However, shortfalls in indirect-cost recovery and the requirement of some sponsors for cost-sharing create a financial burden for the recipients of the funds. Such financial losses are generally compensated for by the gains in intellectual capital that result from greater scientific sophistication, increased academic prestige, more numerous publications, and sometimes patents, which can produce additional income. In sum, research in most AHCs is heavily subsidized by clinical income, which is vulnerable to policies that reduce the revenue from patient care.

The research mission of AHCs has contributed significantly to America's pre-eminence in medicine and biomedical science, but the landscape is changing fast, and the future of research at AHCs is, at best, uncertain. Radical change occurred in 1990 when managed care started to replace the medical faculty's traditional fee-for-service operation; competition from health-maintenance organizations for patients now threatens income flow to AHCs. AHC administrators are scrambling to reorganize their hospital and clinical services and are attempting to establish their own networks of clinical specialists to compete in the primary-care market.
Mergers, acquisitions, and joint ventures with various health-care providers are now common. Such maneuvers are accomplished, however, at the expense of specialty care and of graduate medical education.

It is not yet clear how the new arrangements will affect biomedical research and education, which principally have been conducted by doctors whose salaries were partly subsidized from patient-care income. More than ever, the faculty engaged in research will be expected to fund most, if not all, of their salary, as well as their laboratory costs, from their own research grants. This change is coming at a time when grants are harder than ever to get. In some AHCs, the basic-research enterprise is already being reduced as faculty leave or retire. One can reasonably expect the current stringent conditions will shrink the research enterprise at most AHCs. Moreover, the net impact of managed care is likely to be a devaluation of research success as a criterion for promotion and reward in most medical schools. Without cutting-edge research and a strong academic environment, progress in medical research could languish. It appears that the remarkable era of the traditional AHC is ending, but the full impact of this sea change on the management, philosophy, and morale of medical-school faculties has yet to be realized.

At the same time, financial support of research from pharmaceutical companies has increased substantially in recent years and makes up some part of the support lost because of changes in clinical-practice income.

CHANGES IN RESEARCH AND INSTRUCTION DEALING WITH AGRICULTURE AND NATURAL RESOURCES

Public policies affecting agriculture and forestry were designed to enhance the productivity of US farms and forests. They were focused in particular to enhance the economic status of farmers and to promote general public welfare. The land-grant university system, with its strong components of experimentstation research and extension service, has nurtured an agricultural enterprise that allows the American public to spend a lower percentage of its income for the purchase of food than any other country in the world: between 1956 and 1996, field-crop yields have about tripled while the acreage devoted to agriculture has decreased. The US agricultural research enterprise is therefore perceived by most people to be a bargain.

Over the last 30 years, there has been a serious change in the support of agricultural research. Between 1960 and 1990, the estimated funding for private research in agriculture has tripled; it currently exceeds the investment by both state and federal agencies. These funds have come from chemical, petroleum, and pharmaceutical companies, and a large percentage involves venture capital for biotechnology investments. Although the record of expenditures by companies is not fully disclosed, the sum probably now exceeds $3.5 billion per year. As private investments have increased, there have been major shifts in the kinds of research that are funded. Support of plant breeding has quadrupled and that of animal health has tripled while funds for research on machinery have declined from 36% to 12% of the total invested.

Investments by the states in agricultural research have continued to increase; in sum, they are now much higher than the corresponding federal appropriations. Indeed, the rate of increase in federal support has not kept pace with the needs of teaching institutions. The result has been indirect but negative: a decline in the number of instructional positions that are directly related to agriculture. Many land-grant universities have established programs in molecular biology, biotechnology, sustainable or alternative agriculture, and environmental sciences. Additional changes have been made at some universities to integrate forestry and agricultural research programs, emphasizing studies on regional ecosystems and landscape and wildlife management research programs. The cadre of applied ecologists will need to be increased to cope with these changes in research perspectives.

There is now a pressing need for agricultural-research biologists who are responsive to changing societal requirements to insure the continued availability of agricultural products at a relatively low cost to the consumer while maintaining economic stability for the growers. Such scientists will be essential if we are to provide areas for recreation and ecologic diversity, to conserve and restore damaged ecosystems, and to reduce our dependence on pesticides and other chemicals. Moreover, there will be an ever-increasing need for biologists capable of using the major advances in molecular biology to increase the availability, quality, and safety of food under circumstances that will ensure the sustainability of agriculture and natural resources. The situation suggests that more, not less, should be invested in the agricultural life sciences, broadly defined. The current heavy reliance on funding from the private sector carries some danger that some basic-research problems with less potential for commercial payoff will not get the attention that they need and deserve. That is already evident in the decline of support by major agricultural-chemical companies of research on microbiologic control agents for plant diseases. The emphasis of these companies is on research on and development of transgenic cultivars with disease and insect resistance.

CHANGES FACING INDUSTRY

Before the post-World War II burst of federal funding that created the research-intensive, PhD-granting university, industry was the major supporter of life-science research, and PhDs regularly entered industrial careers. Some mentors and trainees today believe that the only respectable career aspiration is academic research. That opinion is sharply out of phase with the fact that only one-third of PhDs currently obtain academic research positions, whereas jobs in industry have increasingly provided career opportunities for life scientists.

Chapter 3 shows that during the 1980s, when the number of academic research positions was no longer growing rapidly, industry became a major source of jobs in the life sciences. The trends in the 1990s suggest, however, that the growth in the number of industrial research positions might not be as robust in the future as it was in the early 1980s. Several features of industrial organization and patterns of employment are affecting the availability of careers in the life sciences, as discussed briefly below.

DOING THE MOST WITH THE FEWEST

The number of jobs for doctoral-level

microbiologists is projected to grow at an annual rate of 6%; about 15% of the growth represents hiring of postdoctoral fellows, not scientists with permanent positions, according to a recently completed survey by the American Society for Microbiology (Van Ryzin and others 1996). The ASM survey showed, however, that the fastest growth was in emerging fields of biotechnology, such as bioremediation, molecular immunology, and antimicrobial chemotherapy. For some pharmaceutical companies, the highest level of new hiring is in such fields as drug formulation. Chemistry and toxicology show a steady rate of hiring that primarily reflects attrition, with few new positions appearing. By comparison, fields like molecular biology, which saw strong growth in the middle 1980s, are showing no further growth in the 1990s, and replacement hiring might shift toward other life-science disciplines. The ASM survey showed that 57% of industrial respondents forecast increased hiring, but these companies also told the surveyors that future employees must be more flexible and less specialized than their predecessors. At one leading pharmaceutical firm, an increasing number of open positions that were once filled with scientists trained at the bachelor's and master's level are being refilled with PhD scientists.

MERGERS AND OUTSOURCING

In the pharmaceutical and biotechnology industries, the late 1980s and 1990s saw a steady stream of consolidations that resulted in substantial corporate savings with a concomitant disappearance of research positions. The large number of experienced researchers who are therefore on the job market has made it difficult for new PhDs to compete for open positions. In addition, many activities that used to consume large amounts of research time (such as peptide and oligonucleotide synthesis, protein and nucleic acid sequencing, monoclonal and polyclonal antibody production, and receptor-binding assays and immunoassays) have become sufficiently routine that robotics and automation are useful options. Further efficiencies of scale have come from the emergence of new companies that provide the services to pharmaceutical and biotechnology enterprises, but the new positions at these service companies simply offset some positions lost elsewhere in industrial research.

APPLIED VS. FUNDAMENTAL RESEARCH

During the rise of biotechnology in the 1980s, fundamental research was a major part of the work being done by the scientists in the new positions. However, the nature of the industrial research positions has now shifted. The emphasis is now on transforming the fundamental discoveries of the 1970s and 1980s into commercial uses and applications.

Industry continues to down-size, consolidate, and become more efficient. The total volume of industrial research will probably continue to increase but this research is for the most part focused on applied research that has short-term commercial payoffs. Moreover, research on agriculture-related topics is constrained by the commercial value of discoveries. Unlike products with commercial medical applications–whose cost has not, until recently, been prohibitive to development–agricultural research for commercial development is often constrained by the cost of the potential products. Consumers are not willing to pay as much for agricultural innovation as they have been for medical advances; the kind of research that can profitably be pursued in the commercial sector of agricultural research has thereby been constrained.

Even when one understands the economics of a given branch of industrial science, it is generally hard to use the knowledge to predict where increased workforce needs will emerge. Very few people predicted the dramatic emergence of biotechnology before the 1980s.

New fields of industrial research that increase the demand for life-science researchers might emerge. It must be remembered, though, that just as automation and increased efficiency have come along in biotechnology research (for example, in DNA sequencing), technologic innovations that substantially reduce the demand for PhD researchers can be expected to change the patterns of employment of newly trained life scientists.

TRENDS IN GOVERNMENT

As pointed out in chapter 3, the overall fraction of recent PhDs who are employed in government is decreasing, particularly in the older cohorts. If current trends toward government down-sizing and budget balancing continue, federal employment of research scientists cannot be expected to increase. Some growth can be expected, though, in selected fields that are not research-intensive. For example, as reported by Katterman (1996), the number of biotechnology-patent applications filed in the United States has grown about 10% per year since 1990. As more and more genetically engineered products near the marketplace, there will probably be new employment opportunities for life-science PhDs in federal patent-licensing offices and in some regulatory agencies, such as the Food and Drug Administration.

THE DIVERSITY AND SPECTRUM OF CAREERS FOR LIFE-SCIENCE PhDs

ACADEMIC-CAREER TRENDS

Life-science PhDs who seek academic careers with a greater emphasis on teaching might find satisfying careers at several kinds of non-PhD-granting institutions: conventional 4-year liberal-arts colleges that award bachelor's and sometimes master's degrees, 2-year junior and community colleges whose degree is usually an associate in arts, and public and private elementary and secondary schools. An analysis of current employment patterns shows that PhDs are more likely to be found in the 4-year colleges, less likely in community colleges, and comparatively rarely (but not totally absent) on secondary-school science faculties. As the present crop of life-science PhDs in postdoctoral positions seek more permanent jobs, these employment patterns might change, so it is important to examine the current situation with some care.

Comprehensive Bachelors and Masters Degree Granting Institutions

About 20% of the life scientists who are tenured or on the tenure track are now teaching at the roughly 1,150 4-year colleges or universities that do not offer the PhD. These institutions have grown greatly over the last 3 decades, and they have been an important source of employment for recent PhD recipients. Unlike the situation at PhD-granting institutions, the number of faculty positions at 4-year non-PhD-granting institutions has continued to rise, and the number of positions held by life scientists within 10 years of receipt of the PhD increased in both 1993 and 1995 after a period of decline. Because of high student interest in biology as a major, as well as the common focus on preparation for medical school, many life-science departments have grown over the last decade; this trend might continue as students who make up the "echo" of the baby boom matriculate in college. The US Department of Education projects an increase of 0.7 million students in 4-year institutions during the next decade. Assuming that teacher:student ratios remain constant and that there are no changes in instructional practices that might diminish labor requirements, these trends could lead to an increase in the number of life-science faculty.

Most of the biology departments in these colleges are staffed by PhDs who are well trained in research, and most of the faculty are expected to conduct research that employs and trains students. The leading liberal-arts institutions are well known as the source of some of the best graduate students at the top research universities, and it is the research opportunities that they had as undergraduates that prepared these students so well for graduate education. A few such institutions also offer the master's degree. Faculty members have opportunities to pursue their own research interests, but most liberal-arts college professors still spend the majority of their working time instructing students. Salaries at liberal-arts colleges are on the average near or only slightly below those at research universities, but the best-paid teachers at these 4-year institutions are better compensated than those at low-paying universities.

Because most life-science PhDs and postdoctoral fellows have concentrated intensively on research, they have comparatively little experience in teaching, and their qualifications might not be attractive to teaching-intensive colleges. Some graduate students can take advantage of new programs at a number of PhD-granting institutions that offer students exposure to teaching in a more rigorous manner. A small number of "teaching postdoctoral fellowships" have also been developed. One such program (funded by a private foundation) was described to the committee at its public hearing; it provides postdoctoral trainees with 2 years of teaching experience supervised by a mentor. Such a program seems likely to be effective in preparing participants for positions at teaching-intensive institutions.

Two-Year and Community Colleges

The committee found that the 1,471 institutions at this level of higher education employ only about 600 PhDs in life sciences, and the prospects for substantially increasing this number appear to be small. There might be an increased demand during the coming decade, fueled again by the echo generation of the baby boom, which is predicted to increase enrollment at 2-year colleges by about 11%. The impact will probably be quite selective, in that it is apparent that many, perhaps most, of the 2-year institutions do not have a PhD in the life sciences among their faculties.

Secondary Schools

Hiring projections in the COSEPUP report (COSEPUP 1995) suggested that the echo of the baby boom could lead to numerous new positions for K-12 teachers, providing alternative career opportunities for science and engineering PhDs. Our committee believes that this change will probably create a demand for PhDs only at the secondary-school level, and even here the demand is likely to be small. About 0.5% of PhDs in the life sciences are currently K-12 teachers. At that rate, one might expect that 35-40 of the roughly 7,500 PhD's graduating per year would enter precollege teaching. If the rate of entry into secondary schools triples owing to increases in the student populations and increased enthusiasm for the life sciences, the number of PhD life scientists that could be absorbed would be only somewhat more than 100 per year. That is less the 2% of the current production of life-science PhDs so this source of jobs is not likely to have a major impact on career patterns for life scientists.

There are, furthermore, obstacles to the employment of PhD scientists in secondary schools, notably the low salaries and the teacher-certification requirements. Although pay scales for secondary teachers with PhDs are normally higher than for teachers with bachelor's or master's degrees, they are generally lower than the salaries for entry-level assistant professors. Scientists at the end of a 5-12 year period of postbaccalaureate training might well regard secondary-school teaching as a bad bargain. In addition, most states require credentials for a teaching certificate that would necessitate a year or more of additional training in education–also an unappealing lengthening of prejob training. Although a few states have special programs to train candidates with advanced degrees for public-school teaching, the burdens of supporting oneself and paying for this additional training are likely to be serious disincentives. Finally, experienced administrators in public-school systems have offered the opinion that life scientists who are extensively trained in cutting-edge research would not find school teaching captivating.

TRENDS IN LAW, JOURNALISM, AND OTHER FIELDS

With the increase in biotechnology patents and an upsurge in the use of molecular biology as a tool in criminal investigation there has been an increase in the opportunities for life-science PhDs to enter the legal profession. The patent field appears to be dominated by about a dozen large and medium-sized firms. Estimates made in 1997 by patent lawyers at two of those institutions indicate that 20-100 new jobs would become available per year for life-science PhDs. It is customary for PhDs who begin working at law firms to go to law school at night for 3-4 years to earn the law degree that is deemed a necessary credential. Some large firms have clerkship programs that cover law-school costs in exchange for a commitment to continue working for the firms. There is a recent trend to hire PhDs, rather than master's-level scientists, for these jobs because of the large number of highly qualified candidates. PhDs also add to a firm's reputation.

There is a growing interest in journalism among life-science PhDs. Such opportunities appear to be largely associated with the numerous scientific journals that are published, rather than with the more limited number of publishers who handle scientific books. A few life-science PhDs currently working in publishing whom we spoke with thought that future opportunities in the field would probably be constant or perhaps increase slightly. However, competition for careers in journalism is often high. For example, one journalist with a recent PhD in life science moved from a highly regarded specialty journal to a more general publication. There were about 200 applicants for the latter position, and about 50 applications were received for the position vacated at the more specialized journal. Not all the applicants were PhDs, but a doctorate and journalistic experience would appear to have provided the best credentials. The Internet was cited as a medium with particularly good growth potential for scientific journalism.

Some life scientists find positions with private foundations and various other scientific concerns. Again, the competition for such positions is steep. A former assistant professor in the life sciences reported to the committee that there were more than 200 applicants for her present position managing the research-grants program of a philanthropic organization. That figure and others mentioned earlier indicate that there is considerable interest in nontraditional career paths among life scientists. Most PhD programs do not, however, offer the broader exposure and training that would be helpful for entering nontraditional career. The question of whether life-sciences PhD programs should change to offer this additional training is addressed in chapter 6.

In summary, our findings suggest that the number of positions in nontraditional fields of employment for life-science PhDs appears to be rather small, and that the competition for these jobs is strong. The committee acknowledges that it cannot predict the emergence of entirely new employment opportunities that might change employment characteristics considerably. Several sites on the World Wide Web (for example, Next Wave: An Electronic Network for Young Scientists) offer career information that might be of interest, and appendix G contains a list of Web sites that provide data and career information for life scientists.

REFERENCES

Clotfelter CT. 1996. Cost escalation in elite higher education. National Bureau of Economic Research monograph. Princeton, NJ: Princeton University Press.

COSEPUP (National Academy of Sciences, Committee on Science, Engineering, and Public Policy). 1995. Reshaping the graduate education of scientists and engineers. Washington, DC: National Academy Press

Jones R and Sanderson S. 1996. clinical revenue used to support the academic mission of medical schools: 1992-1993. Academic Med 71:3

Katterman L. 1996. Biotechnology patent boom offers career opportunities for scientists. The Scientist 10:15-6.

Ruzek JY, O'Neil EO, Williard RL, Rimel RW. 1996. Trends in US funding for biomedical research. San Francisco: UCSF Center for the Health Professions.

Van Ryzin G, Dietz S, Weiner J, Wright D. 1995. The employment outlook in the microbiological sciences, 1995. http://www.asmusa.org/pasrc/empoutlk.pdf.

5 IMPLICATIONS OF THE FINDINGS

CHANGING CAREER PROSPECTS FOR LIFE-SCIENCE PHDS

The career prospects in 1998 for a graduate student or postdoctoral fellow in the life sciences are very different from those of someone who trained in the 1960s or 1970s. Today's life scientist will commonly have started graduate school at a slightly greater age and will have taken 2 years longer to obtain the PhD degree. This year's PhD recipient is on the average 32 years old. With degree in hand, he or she will probably join an ever-growing pool of postdoctoral fellows now estimated at about 20,000 persons to engage in research while obtaining further professional training. Although postdoctoral positions have much in common with medical internships and legal clerkships as a means to obtain further postgraduate training, they are different in one important respect: they have no fixed length of tenure. It is not unusual for a trainee to spend 5 years or more as a postdoctoral fellow. Consequently, the average life scientist will be 35-40 years old before obtaining his or her first permanent job.

A life scientist's probability of finding employment in either a 4-year undergraduate college or a research university has declined over the last 20 years, as described in chapter 3. In contrast to declining prospects in academe, however, the fraction of graduates who hold positions in industry has increased; it surged during the middle 1980s, but the increase has slowed recently. In spite of the increase, according to the National Research Council surveys, there has been an overall decline in the percentage of life scientists who are using their research training in their "permanent" employment; the fraction of life scientists who had graduated 5-6 years before and who were employed in "permanent" positions in academe, industry, or government decreased from 89% in the 1973 survey to 62% in the 1995 survey[1].

CHANGES IN THE RESEARCH AND TRAINING ENTERPRISE

The rapid expansion in federal support of basic biologic research that occurred during the 1960s and early 1970s allowed the joint research and training system to flourish. Scientists who earned their PhDs in that era had bright prospects for employment in research. The training system of that time was built on the tacit premise that there would be continuous growth in the size of the US research enterprise–sufficient to absorb the trainees who were moving through the system. The result was not simply that more life scientists were available to work in laboratories and in the field; the active training enterprise produced a scientific workforce whose age distribution became skewed toward youth. That age bias brought energy and innovation into the profession.

Beginning in the early 1970s, however, the rate of expansion in federal research support and the growth in the number of universities and colleges began to slow down. The slowdown was not accompanied by a corresponding decline in PhD production. Instead, the annual rate of PhD

---

[1] See Figures 3.12 and 3.13, in chapter 3. The categories included as employed in "permanent" positions are tenured or tenure-track faculty positions in PhD-granting or other academic institutions, positions in industry or government, and other positions including self-employment. The categories included as not employed in "permanent" positions are unemployed and seeking a position, part-time employment, positions outside science and engineering, postdoctoral appointments in any sector, and other academic positions.

production was fairly constant through the 1970s and 1980s at about 5,500 per year. Two changes in the employment market absorbed the trainees who could no longer find jobs in the traditional employment sectors of academe, the pharmaceutical and agricultural industries, and government. First, the biotechnology industry emerged in time to provide new and exciting employment prospects for many PhD graduates in the life sciences. Second, the system adapted to the continued high rate of training by increasing the support available for postdoctoral fellows.

The resulting expansion of the postdoctoral pool has not, however, created permanent jobs for life scientists; it has produced a holding pattern. In its favor, the increased fraction of PhDs who now take postdoctoral work is probably responsible for the finding that an increased fraction of life-science PhD recipients are involved primarily in research (Table F.1). The result has been an economical and highly effective workforce whose research productivity is excellent and whose salary costs are comparatively low. The intellectual fluidity and scientific productivity of the life sciences rests to a great extent upon this cadre of postdoctoral fellows who, with graduate students, operate within the tradition of laboratories that are funded through highly competitive grants to principal investigators for the pursuit of their scientific ideas.

If the annual rate of PhD production had been constant into the 1990s, the number of scientists in the postdoctoral holding pattern would probably have continued to grow. In reality the rate of PhD production has increased. In 1996, 7,696 life-science PhD degrees were awarded, roughly a 42% increase over the 5,500 characteristic of the 1980s. A substantial fraction of that increase was due to an influx of foreign students, partly as a result of a change in immigration law described in chapter 2. In 1995 about 22.4% of the PhD recipients were foreign nationals. Although it is difficult to know precisely what percentage of those foreign-born graduates will return to their countries of origin, the most recent Survey of Doctoral Recipients indicates that, at least at graduation, the majority state an intention to remain in the United States.

The dramatic increase in the number of life-science PhDs has already had a substantial effect on the size and composition of the postdoctoral pool, and the pool is being enlarged by an influx of foreign-trained PhDs who have come to the United States for further training. The inevitable consequence has been an increase in the competition among postdoctoral fellows for permanent positions in all employment sectors. The full impact of the population increase has not yet been felt in that most of the new postdoctoral fellows have yet to face the permanent-job market. That suggests that young people's difficulty in finding jobs that use their research training will get worse before they get better. Moreover, the committee's analysis in chapter 4 suggests that there is no new source of jobs for life scientists lying just over the immediate horizon–nothing like the opportunities provided by industry during the 1980s. If anything, the expected changes in the financing of higher education, academic health centers, and industry will only widen the gap between the number of life scientists being trained and the number of jobs for them to do.

IS THERE A PROBLEM? AN ANALYSIS FROM DIFFERENT PERSPECTIVES

Should the recent changes in the career paths of life scientists be a cause of concern? Is the dismay that is being voiced by the current generation of trainees a symptom that the system is no longer optimal, or is it simply the normal discomfort of students reacting to the prospect of healthy competition? Opinions about the value, appropriateness, and stability of the current professional system vary widely, depending in part on the perspectives of those holding the opinions. A convenient way to describe the situation is to identify groups of "stakeholders" who look at the current professional system from different points of view.

ADMINISTRATORS AND ESTABLISHED RESEARCHERS

Leaders of industrial or government laboratories, university administrators, teachers in large undergraduate programs where extensive laboratory work is performed, and established life-science researchers who must compete for renewed funding are likely to argue that the current situation has much to offer; their motivation to promote change is weak or absent. Both the time-consuming experiments that are characteristic of much biologic research and the education of large numbers of undergraduates are well suited to the skills and training of graduate students and postdoctoral fellows. The research productivity of an individual laboratory–even of an entire department–can depend on the number of graduate students employed, so future funding and intellectual prestige might depend on attracting as many good students as possible. Occasionally, there are additional incentives to keep numbers of students high, such as the supplements provided by some local legislatures to their state universities in proportion to the size of their graduate programs. All those factors are powerful arguments for leaving the current situation unchanged.

Few branches of the life sciences in the United States have adopted the alternative professional system of hiring permanent laboratory scientists and technicians trained at the bachelor's, master's, or PhD level. From an economic point of view, such permanent employees usually require higher salaries and a greater institutional commitment, such as retirement benefits, than temporary students and fellows. Furthermore, from an intellectual perspective, most life scientists will argue that students and postdoctoral fellows bring fresh approaches and new energy to a laboratory–features that are difficult to duplicate with a more permanent workforce. Thus, a pool of young scientists who rotate through a research laboratory is considered by many to be optimal for creativity and productivity, even though there can be inefficiencies while students are acquiring expertise.

FUNDING AGENCIES

Organizations that fund life-science research can also be seen as having a vested interest in maintaining the status quo. Life-science graduate students supported by research grants are regarded by many such agencies as employees, as reflected by their designation on budget sheets and the resistance of some agencies to paying tuition. Most life-science graduate students are good value for the research dollar: they earn annual salaries of only about $16,000 and generally work very hard. Their productivity might be modest early in their doctoral research, but they become effective producers of data later in their training. In this context, it appears that a long graduate-student tenure has features that are desirable to established scientists and funding agencies; this training system increases the likelihood that a student can accomplish substantial work while still being paid at a comparatively low rate.

Funding agencies are likely to view their investment in postdoctoral fellows in much the same light. Even though the initial salaries of this group are higher than those of graduate students, tuition is no longer an issue, and these young scientists are more likely than graduate students to be immediately effective research workers. Thus, the growth of both populations of life scientists carries benefits for institutions that wish to maximize the effect of their research investment.

INCOMING GRADUATE STUDENTS

Prospective graduate students have good reasons for wanting the profession to maintain high enrollments in a large number of graduate programs. The availability of many programs offers students a wide range of choices, and high enrollments increase one's likelihood of being accepted. Stipends for graduate life-science students are below the current average starting salary for a person with a bachelor's degree in biology ($21,558), so short-term financial sacrifices are associated with graduate training, but one can reasonably expect to recover these losses eventually. Finally biology has an exciting intellectual future, and students can be confident that the research apparatus will not run out of work in the foreseeable future.

SENIOR GRADUATE STUDENTS

Senior graduate students might begin to view the current training system more negatively. The data show that they must expect a protracted graduate career; the longer their training continues, the greater the extent to which their incomes will fall behind the salaries of their college classmates who entered the workforce at graduation. Health-insurance benefits might not be as good as those in the overall workforce–a more pressing issue as a student contemplates starting a family. During the later stages of training, senior graduate students might no longer be learning new skills but rather spending time in increasing their professional accomplishments and contributing to those of their mentors.

POSTDOCTORAL FELLOWS

Finding a postdoctoral position is normally not difficult because many such jobs are available. The compensation of life-science postdoctoral fellows is, however, only marginally better than that of graduate students, and the quality of the benefits remains low. At the beginning of this career stage, postdoctoral fellows might well be so involved with their new and exciting work that their long-range professional prospects are invisible. Virtually all by their third or fourth year, and some sooner, face the prospect of searching for a more permanent position. Many entered graduate school with the intent of eventually finding a position as a professor in a university or college. Their mentors in both graduate school and postdoctoral training probably encouraged them to pursue this career goal, and some will have implied, either explicitly or implicitly, that any other career outcome would be a sign that they had failed. Yet the likelihood that they will obtain such a position is now lower than it was when they made the decision to begin graduate studies. Although unemployment is very low (still less than 2% in Table F.1) and underemployment is only modest, the number of applicants for good jobs of all kinds–whether in academe, government or industry–is very large. Thus, the prospects for permanent employment that will provide research opportunities and intellectual independence appear dim.

Even the most highly successful postdoctoral fellows, working in one of the 26 institutions of the highest reputation, are now seeing that 3-4 years of postdoctoral training might not be sufficient to secure a good job. The data in Table F.1 show that the fraction of scientists in the cohort 3-4 years after receipt of the PhD who are still engaged in postdoctoral training has been steadily increasing over the last 10 years. Members of that cohort are competing for jobs with members of the cohort who are 5-6 years post receipt of the PhD, who have often published more papers. In response to these realities, many postdoctoral fellows are now undergoing a "crisis of expectation" that comes from a sense that an implicit contract between them and the scientific establishment has been broken. They had agreed to forgo economic compensation for 10-12 years while they acquired scientific knowledge and expertise; in exchange, they expected a reasonable likelihood of obtaining a satisfying job later. Had they known their realistic prospects at the beginning of the long training period, they might well have made different choices.

YOUNG INVESTIGATORS

Another important group of stakeholders is the young scientists who have recently become employed in research-oriented institutions. One might imagine that they would view their careers as established and that they would adopt the viewpoint of more-senior scientists. Several differences between young and established scientists, however, suggest otherwise. For one thing, these scientists are likely to be older than were life scientists at a comparable stage of professional development some years ago. The demanding work of establishing a productive laboratory comes at a time when other responsibilities, such as children, might be competing for their time. Decisions about starting a family are important to both male and female students, but females must consider whether they want to have children because they are likely to be in their middle to late 30s, and their biologic clocks will not grant them much more time.

Young life scientists whose jobs are not in an industrial or government laboratory face the primary responsibility of attracting research support so that they can build their research programs and have some likelihood of being retained and promoted. They must compete successfully for money, or their research careers will soon end. Yet success rates in obtaining grants have decreased for young investigators as they have for investigators of all ages. The situation has been ameliorated to some extent by the existence of other sources of research money that are available explicitly for young people, such as grants from the Pew Charitable Trusts, the Searle Foundation, formerly from the Markey Trust, and now from both the Burroughs Wellcome Fund, and the American Cancer Society, which is focusing its scientific-grants program on young people. Notwithstanding the additional sources, however, even the most successful young investigators view the task of establishing their research programs as stressful and difficult.

THE AMERICAN PEOPLE

An additional group of stakeholders is the American people, the citizens whose taxes and gifts have supported all aspects of the scientific enterprise. The American people have a right to expect a system of life-science research that will be productive and efficient and that will generate knowledge that leads to improvements in their environment, their food, and their health.

Through Congress, the electorate has consistently endorsed the importance of life-science research, and such groups as Research!America have found that most Americans are willing even to increase the money invested in biomedical research (Research!America 1997). From an economic point of view, there is much value in the short run associated with a large training enterprise that keeps labor costs low, but this might not be the most cost-effective strategy to meet the research interests of the country in the long run. Taxpayers deserve a professional system that will be strong and effective not just today, but also in the future. The interests of the American people will be best served by keeping firmly in mind the question of what is best for life-science research enterprise, not just best for some current life scientists.

THE CRISIS OF EXPECTATION

The foregoing discussion underscores the reality that one's opinion about the fairness and effectiveness of the current system for producing life scientists and conducting life-science research can depend very much on how far along the career path one is. Many established scientists view the current professional system as optimal and point out the importance of competition for a healthy scientific climate; these scientists often refer to their own success with analogous competition when they were young. There is certainly some truth in that point of view, but it misses some of the flavor of the current times. The current cohort of established investigators began their careers in a very different climate; regardless of their recollections, they experienced far more favorable conditions—from the length of their training to their prospects of a job and a grant with which to conduct research.

The crisis of expectation among today's young life scientists is palpable. Although there are no extensive data from an objective survey of public opinion, the committee had information from four informal sources. In the fall of 1994, Richard McIntosh, president of the American Society for Cell Biology, wrote a short piece in the society's newsletter (Mcintosh 1994) describing his understanding of the problems facing young cell biologists and asking those interested to reply and present their views or experience. More than 50 letters were received; some were written by senior investigators, but most came from graduate students, postdoctoral fellows, and young independent scientists. More recently, the committee held a public hearing in Washington and invited members of the life-science community to present their views at the hearing and electronically through e-mail. The committee was also given access to the results of a survey conducted by the University of California, San Francisco Center for the Health Professions of the Pew Scholars in the Biomedical Sciences. This program, funded by the Pew Charitable Trusts, has supported 20-22 newly independent scientists per year for the last 10 years. Pew scholars are a highly select group of young investigators in all fields of the biomedical sciences. The survey collected retrospective data on the duration of training and opinions of the scholars regarding the health of biology. Finally, the Education Committee of the American Society for Cell Biology, chaired by Professor Frank Solomon of the Massachusetts Institute of Technology, used a Federation of American Societies for Experimental Biology e-mail network to query a broad range of investigators about their views.

Clearly, those informal surveys cannot be regarded as statistically reliable inasmuch as no effort was made to obtain a representative sample of the various populations of life scientists. Nonetheless, they are informative in several ways. First, they encourage the view that many established scientists are concerned about the fate of the young people they are training, many of whom are having great trouble getting jobs or grants. Second, there is a perception that a large gap separates the haves and the have-nots: those who are established in jobs and with grants and those who aspire to such a situation. Third, there is a pervasive sense that in the current climate of increased competition, something precious has been lost; the excitement and promise that have characterized the life sciences for many years are not felt with the same intensity by many young people because they are too concerned about their futures. Fourth, there is a widespread sense of failed expectations. Most of the young people who replied had entered life-science training with the expectation that they would become like their mentors: they would be able to establish a laboratory (in industry, academe, or a government agency) in which they would pursue research based on their own scientific ideas. The reality that now lies before them seems very different. There simply are too few such jobs, in any sector of the profession, to hire all the new life-science aspirants of high quality. The result is a crisis of expectations.

Implications of the Findings

Many thoughtful commentators on the current situation, including the National Academy of Sciences' Committee on Science, Engineering, and Public Policy report (COSEPUP 1995), have argued that there are plentiful alternative careers for people with the intellectual abilities and training implied by a doctorate in the life sciences. Whether or not those positions will become more important as sources of employment for life-science PhDs in the years ahead, there appears to be a substantial resistance to career redirection during the postdoctoral years. At least four factors seem to contribute to this unwillingness to redirect a career:

- Most people who have gone through the labor of getting a life-science PhD, whether or not they go on to training at the postdoctoral level, love the process of science in a powerful and fundamental way. To relinquish the pursuit of a first professional love is a tremendous loss.
- It is satisfying and rewarding to do something that one does well. Most PhD-trained life scientists are highly accomplished in their research, and there is intrinsic satisfaction in doing more of same.
- The expectations with which many people entered scientific training included working in a field that is highly respected within the country, earning a good middle-class wage and doing things that are fundamentally enjoyable. These are attractive features of life-science research; leaving science before one is forced out is therefore very difficult.
- When one has invested so much effort in highly focused training, it seems wasteful and even self-destructive to leave it behind and go on to something else. There are transferable skills—such as problem-solving, the acquisition and analysis of data, and the hierarchic organization of ideas and activities—but many postdoctoral scientists expect that a change of fields will mean either doing something rote or going through yet more training. After more than 10 years of "training", this is an onerous prospect.

FACTORS AFFECTING THE FUTURE VITALITY OF THE LIFE-SCIENCE ENTERPRISE

One important aspect of America's current training system for life scientists is beyond dispute: it is inherently expansionist and is not at steady state. The significant contributions of young people to the life-science enterprise have made them so attractive to the senior members of the profession that the rates of training have continued to increase while the number of people still in postdoctoral positions, without any immediate prospect of permanent research positions, is also increasing. The most likely future for a recent life-science Phd is to be a postdoctoral fellow for a very long time.

The present situation in life sciences is not, however, unique. All the sciences expanded rapidly in the late 1950s and the 1960s as a direct response to the threats of the Cold War. The number of academic openings was huge, coming from both expansion in existing universities and the rapid creation of new ones. That growth was highly unusual in the history of science, and it is unlikely to be repeated soon. As the inevitable slowdown occurred, there developed an over-abundance of aspirants relative to the number of permanent positions in the sciences. In physics, the reduction in research funding reduced both available positions and funds to support research and training; as a consequence, enrollments in physics programs declined.

The effect of the slowdown was felt earlier in fields other than the life sciences, in part because the life sciences have experienced a virtual explosion in opportunities and their federal support over the last 10 years has outperformed that of all other sciences. In addition, the life sciences have made efficient and effective use of the postdoctoral position by keeping remuneration of younger colleagues low. As a consequence, the life sciences have been able to support a much larger number of postdoctoral fellows than any of the other sciences.

The current pressing challenge for the community of life scientists is to acknowledge that the *structure* of the profession has led to declining prospects for its young and to develop accommodations that maximize the quantity and quality of future scientific productivity. Success in meeting the challenge will depend to a large extent on ensuring the future success of the most talented of young life scientists. In the next section of this chapter, the committee analyzes the effects of the structural changes from the perspective of the scientific enterprise itself.

NUMBER OF ASPIRANTS

The current size of the life-science PhD candidate pool is testimony to the remarkable success of the US investment in life-science research over the last 20 years. Many college-age students, both here and abroad, judge the life sciences to have the most exciting future of all the sciences. As a result, the enrollment in undergraduate life science courses is growing: from 1989 to 1993, the number of people earning bachelor's degrees in the life sciences increased by about 30% (NSF 1996). The future vigor of the life sciences will depend on ensuring that the most talented students continue to be attracted to graduate training in the life sciences. Of course, the fascinating problems that remain to be solved will always be a draw, but to provide these able young people a profession that is commensurate with their talents we must meet at least two additional conditions: we must inform them in realistic terms of their chances of achieving their career goals and we must recognize that these times are very different from those when today's established investigator began their careers. Several of the recommendations presented in chapter 6 focus on meeting those conditions effectively.

BALANCE BETWEEN RESEARCH TRAINING AND EMPLOYMENT OPPORTUNITIES

The extraordinary research opportunities that are sketched in chapter 4 are only a few of the many in modern life science that offer stimulating challenges for both scientific advance and commercial development. As a reflection of the scientific opportunities, the budget of the National Institutes of Health (NIH) has fared exceptionally well in Congress over the last 10 years, when other discretionary programs of the federal budget have diminished. The FY 1997 budget included a remarkable 7% increase for NIH–unprecedented among agencies funded within the discretionary part of the budget. That vote of confidence on the part of the president and Congress reflects their conviction that the life sciences are important to the future health and economic well-being of the US population.

In the context of the scientific and financial opportunities there appears to be no compelling justification for discouraging the best students from considering graduate training in the life sciences. As long as there are numerous tasks to be done and sufficient funds to support research, the training of new scientists has a high priority for the profession. Moreover, the long time between entry into graduate school and assumption of a permanent position makes it difficult to predict the employment market as little as 10 years hence.

But it would be irresponsible to ignore the signs that our existing PhD production is perhaps too large and that there is an imbalance in the population of life scientists compared to available positions. The signs include the lengthening of time to graduate-degree receipt and the increases in the duration and number of postdoctoral positions. It is argued by some that the lengthening of training reflects the vast amount of new information that must be learned to become a successful modern biologist, but this argument is difficult to sustain on either intellectual or practical grounds. As knowledge increases, some of what used to be thought essential is set aside, and more of what is still essential is taught at lower levels. High-school students now learn about the structure and function of DNA, whereas 30 years ago this was college material. The committee believes that the lengthening of graduate and postdoctoral training is primarily a response to the growing number of applicants and the intense competition for permanent positions. To be competitive for those positions, young scientists must have extensive records of productivity at each stage of their careers.

The continued increase in graduate admissions over the last 10 years has contributed new strains to an already strained system. One can easily imagine that further increases in graduate enrollments, without a concomitant increase in the size of the job market, will lead to such widespread student disaffection that the long-term result will be a drop in the number of highly qualified PhD candidates in the life sciences. The situation suggests that a balance must be found to maximize the likelihood of a good supply of high-quality, well-trained life scientists for many years to come.

STRATEGIES FOR OPTIMIZING GRADUATE AND POSTDOCTORAL TRAINING

MAXIMIZING THE RETURN ON FUNDS INVESTED IN TRAINING

The stipend and tuition of US-trained graduate students in the life sciences are supported by a variety of mechanisms, as described in chapter 2, including training grants, fellowships, and teaching and research assistantships. About half the students are employed as research assistants. The different sources of support have relatively little effect on the day-to-day activities of students, the vast majority of whom spend their time conducting research in the laboratories of their mentors. However, there is a real distinction among the funding mechanisms in the level of oversight of training itself. We focus in the following pages on the NIH support of training because NIH is the single largest source of such support. Other federal agencies play important roles and, as can be seen in Table 2.1, institutional support of graduate students and "other" support, including self-support, also account for substantial numbers of students.

The current NIH training-grants program was established by Congress in 1973 when it authorized National Research Service Awards (NRSAs) as a way to ensure that the need for new biomedical and behavioral research scientists was being met. At the same time, Congress asked the National Research Council to make periodic estimates of the national needs for such personnel that congressional committees could use to evaluate the annual NIH budgetary requests for training funds; this action was intended to prevent shortfalls and surpluses in the number of research scientists being trained. For more than 20 years, the Research Council's Committee on National Needs for Biomedical and Behavioral Research Personnel has been making advice available to Congress.

Training grants are awarded to graduate programs on the basis of a stringent process of peer review. The grants fund the stipends and some fraction of tuition for a specific number of students, determined at the time of application review. Some funds are also provided for auxiliary educational activities, such as seminar programs and symposiums. Graduate students are identified for appointment under a training grant by the institution itself, and they are usually supported for 2-3 years of their total graduate career. NIH supports about 7,500 students on training grants at about 197 institutions, or about

14% of the country's life-science graduate students.

NIH training grant are awarded only after a graduate program has been peer-reviewed by a training committee appointed by the NIH. The review process takes into account such factors as students' time to degree, postgraduation careers, and accomplishments. The process also holds programs to a very high standard of minority-group student recruitment and retention and faculty diversity. And applicant institutions must provide a program of formal instruction in the responsible conduct of research.

The review committee visits the training institution and observes the educational program, interviews students, and engages faculty in discussion. That kind of review by an external group brings to training an expert assessment of quality that parallels the scrutiny that research proposals receive. Such careful examination of faculty, students, and graduates stands in marked contrast with the procedure for employing a graduate student as a research assistant under a research grant, in which case the judgment of the supervising investigator and the willingness of the student are the only controls on the quality of training. In the committee's opinion, the guidance achieved through the review process is likely to produce a better-balanced, more-rounded education of students. Most important, perhaps, is that the award of a training grant is based on the quality of training provided and the training record of the program, and not just on the value or significance of ongoing research. Competition among universities for training grants is fierce. In general, the programs that succeed in obtaining training grants are those in the top-rated universities, as ranked by the National Research Council's Survey of Graduate Programs (NRC 1995)

The superiority of outcomes of training grants is is difficult to document. One older study of the question (IOM 1984) focused on the biomedical sector of the life sciences. The study compared performance with respect to a series of indicators (for example time to degree, completion of degree, later research-grant awards, and articles written) of three groups of former graduate students: those who had held NIH traineeships, others in the same programs who had not had traineeship support, and all other biomedical graduate students in the same annual cohorts. Holders and nonholders of traineeships in programs that had training grants performed about the same, and both outperformed the students who had completed programs that did not have any training grants. It appears that the benefits of training grants are programwide rather than support-specific. The results of that study, which is now 17 years out of date, would appear to support the committee's judgment that applying for and receiving a training grant have a salutary effect on department faculty, leading them to a concern about how, as an entity, they are providing for the education and training of their students. An update of the study is being sponsored by NIH, but its conclusions were not available at the time of our deliberations.

Those results are equivocal in that training grants are awarded only to programs that are already providing a superior education or have attracted students of superior ability. The alternative explanations cannot be ruled out, and the prominence of highly ranked institutions on the roster of those receiving training grants lends them added plausibility. Nevertheless, members of the committee with personal experience of the review process for training grants believe that the process affects the critical standards that faculty apply to themselves. On this ground alone, namely the beneficial scrutiny of peers who are not immediate colleagues, seems to be the strength of NIH training grants.

Almost 12,000, or two-thirds, of the graduate students supported by federal funds in 1995 were

Implications of the Findings

paid from research grants awarded to faculty (see table 2.1). Unlike the training grants and fellowships awarded to individuals, the quality of graduate training provided through this mechanism is not monitored by any agency outside the individual university. NIH, the major federal sponsor of research and training, does not consider the research funds used for graduate-student salaries on research grants as money invested in training although tuition and a salary can be charged to the grant. Rather, these students are seen as employees hired to conduct research. According to Public Health Service policy, graduate students' tuition remission that is charged to faculty research grants is an allowable cost—payment in lieu of salary or wages to students performing necessary work.

Supporting student training through individual research grants permits a funding agency the least amount of peer review of its graduate training investment. It also promotes an employer-employee relationship between faculty mentor and student that creates a potential for a conflict of interest that might adversely impact effective training. For example, because PhD training does not have a fixed term, the decision as to when a candidate has completed training usually rests with one or a small number of faculty members. This system contains a potential for abuse, particularly in times of job shortage. A conflict can arise between a student's interest in moving on to the next career stage and a professor's interest in retaining a highly productive worker. Or a mentor might discourage a student from taking additional coursework or teaching an additional class to gain more pedagogic experience on the grounds that these activities take time away from the grant-supported activity.

NIH and the National Science Foundation also award graduate-training fellowships directly to individuals, although the number of fellows at any time is tiny compared with the numbers of trainees and research assistants. Fellows usually enjoy more freedom in shaping their graduate education than do trainees and assistants, although they must of course abide by department or program rules. In considering fellowship applications, the overall quality of the institution chosen for training is taken into account, but the major factor in awarding a fellowship is the quality of the applicant. Once such a fellowship has been awarded, there is no followup review to judge the nature or quality of the training that the awardee has received. This form of graduate support therefore lacks an important component of peer review that is found in training grants. By relying more on training grants for the support of graduate students, the federal government will be in a better position to gather information about its current investment in graduate education and thus be in a better position to monitor PhD production.

THE PROBLEM OF TIME TO DEGREE

Whether the pressure to lengthen post-baccalaureate training is coming from mentors, who are maximizing the return on their investment in training, or from the students themselves, who are trying to improve their research records, the outcome is that young scientists are spending their most creative and productive years under the direction of more senior investigators. The US scientific enterprise is at risk of losing what many consider to be its most distinctive and successful attribute: that scientists are given their independence at a relatively early age. In contrast with many European countries, where scientists spend many post-PhD years in positions that depend on senior professors, the United States has prided itself on encouraging the energy, independence, and creativity of its talented young practitioners. In the past, it was expected that by the age of 35 US life scientists would have their own laboratories and the resources to carry out newly conceived research plans.

Figure 5.1 and table 5.1 show the number of tenured and tenure-track faculty of various ages at PhD-granting and non-PhD-granting institutions in 1975, 1985, and 1995. The distribution in 1975 was decidedly skewed toward a young faculty complement. By 1994, the distribution was broader and shifted toward higher ages (Figure 5.2). Whereas in 1975, half the faculty were under 39-40 years old, half of the faculty in 1995 were under 47-48.

Although young scientists might be productive in dependent postdoctoral positions, it is important to consider whether they are allowed, under these circumstances, to develop and use their creativity. The lengthening of time that young scientists spend in dependent positions would be deleterious to science only if there were a negative correlation between age and scientific innovation. In mathematics, the aging of the population would be viewed with great dismay, given the common perception that mathematics benefits from young and nimble minds. In the life sciences, there is not the same perception that youth is an advantage. However, using the Nobel prize as a yardstick of originality and impact of scientific work, Stephan and Levin (1993) examined the age at which the critical experiments awarded Nobel prizes in Medicine and Physiology in 1901-1992 were conducted. They found that the median age was 38 years, only slightly older than the median age of 37 in chemistry and 34.5 in physics. Their data showed that the most innovative experiments generally were done by those 30-50 years old; the majority were under 40. The authors concluded that "it is safe to say that regardless of field, the odds of commencing research for which a Nobel prize is awarded decline dramatically after age 40, and very, very few laureates undertake prize-winning work after the age of 55."

Fiugre 5.1 Number of US life-science PhDs in tenured positions, by age, 1975, 1985, 1995.

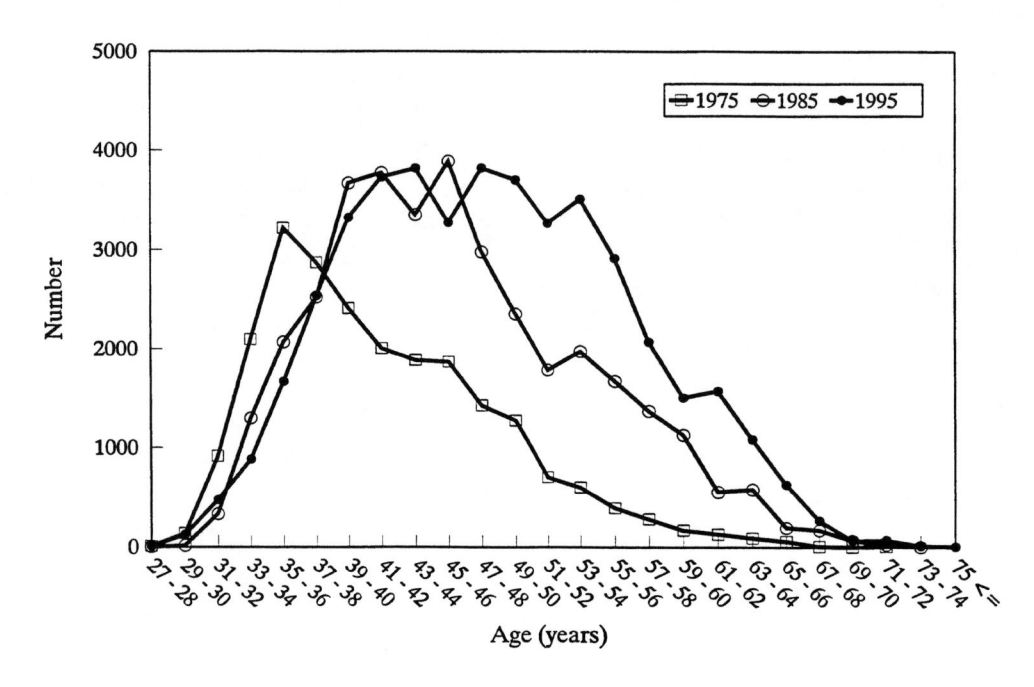

Data from table 5.1.

Table 5.1 Age distribution of US PhD life-science faculty in 1975, 1985, and 1995

Age, Years	1975 Survey No.	1975 %	1975 Cumulative %	1985 Survey No.	1985 %	1985 Cumulative %	1995 Survey No.	1995 %	1995 Cumulative %
27-28	5	0.0		0	--	--	2	0.0	0.0
29-30	132	0.6	0.6	13	0.0	0.0	122	0.3	0.3
31-32	912	4.1	4.7	329	0.9	1.0	471	1.1	1.3
33-34	2093	9.3	14.0	1295	3.6	4.6	881	2.0	3.3
35-36	3218	14.3	28.3	2067	5.8	10.4	1664	3.8	7.1
37-38	2868	12.8	41.1	2523	7.1	17.4	2533	5.7	12.8
39-40	2410	10.7	51.8	3668	10.3	27.7	3324	7.5	20.3
41-42	2002	8.9	60.7	3772	10.6	38.3	3726	8.4	28.7
43-44	1882	8.4	69.1	3353	9.4	47.7	3817	8.6	37.4
45-46	1865	8.3	77.4	3886	10.9	58.5	3274	7.4	44.8
47-48	1421	6.3	83.7	2977	8.3	66.9	3821	8.6	53.4
49-50	1268	5.6	89.3	2353	6.6	73.5	3700	8.4	61.7
51-52	699	3.1	92.4	1782	5.0	78.4	3267	7.4	69.1
53-54	595	2.6	95.1	1971	5.5	84.0	3510	7.9	77.0
55-56	388	1.7	96.8	1668	4.7	88.6	2913	6.6	83.6
57-58	276	1.2	98.1	1366	3.8	92.5	2069	4.7	88.3
59-60	164	0.7	98.8	1124	3.1	95.6	1501	3.4	91.7
61-62	123	0.5	99.3	551	1.5	97.1	1567	3.5	95.2
63-64	86	0.4	99.7	577	1.6	98.8	1079	2.4	97.7
65-66	54	0.2	100.0	190	0.5	99.3	626	1.4	99.1
67-68	6	0.0	100.0	167	0.5	99.8	264	0.6	99.7
69-70	0	0.0	100.0	61	0.2	99.9	61	0.1	99.8
71-72	0	0.0	100.0	24	0.1	100.0	67	0.2	100.0
73-74	0	0.0	100.0	1	0.0	100.0	8	0.0	100.0
75+	0	0.0	100.0	0	0.0	100.0	4	0.0	100.0
	22,467			35,718			44,271		

Those authors attributed the association between important scientific discovery and youthfulness to many factors, including the ability of the young to focus on a problem without the distractions and responsibilities that people accumulate with age. They also identified the ability to approach a problem from a fresh perspective unfettered and unbiased by previous experience and the freedom of having little to lose from being wrong. Today, life scientists are still in dependent positions well into their 30s; often they are working on research projects designed by their mentors rather than on projects that they designed themselves.

It can be argued that the age-related success of Nobel laureates, a highly elite group of scientists, does not reflect the population as a whole. One indication that age does affect the creativity of a broad range of life scientists is the observation that the likelihood of any person's competing successfully for an NIH grant

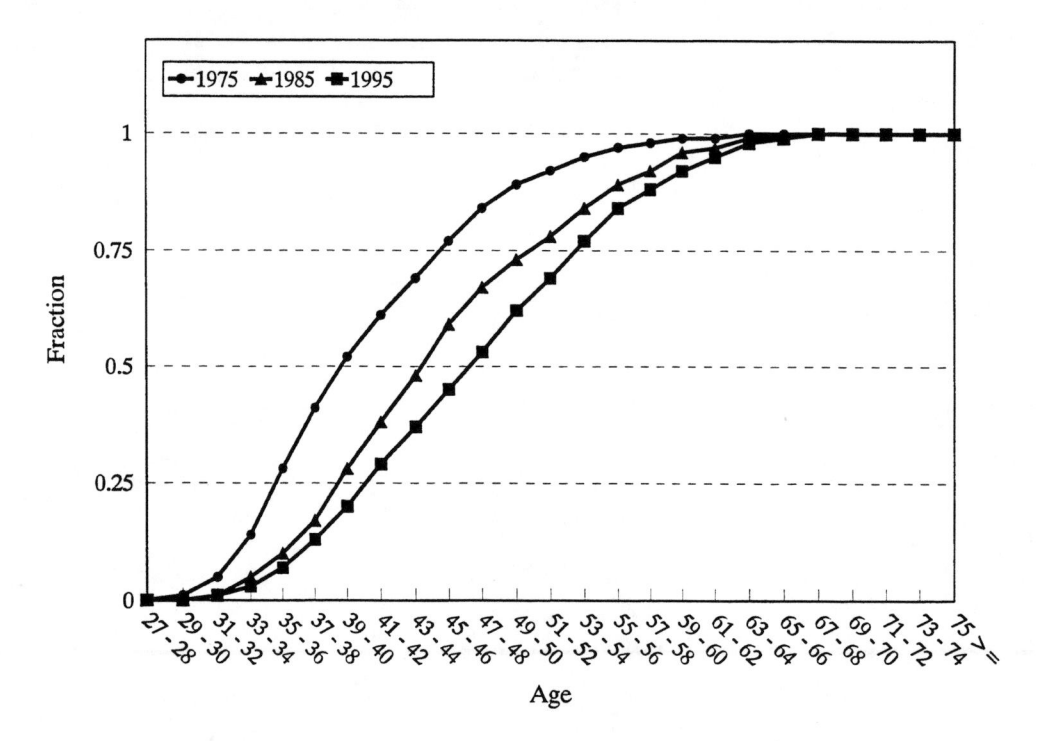

Figure 5.2 Cumulative fraction of US life-science PhDs in tenured positions, by age, 1975, 1985, 1995.

Date from Table 5.1.

decreases after the age of 50. Given that trend, it is reasonable to worry that delaying the independence of young scientists until they are well into their 30s or early 40s, will have long-term deleterious effect on the quality of science produced. Other impediments to the continual replenishment of university and college faculties with young scientists, such as tenure and the disappearance of mandatory retirement because of age, also contribute to the "graying" of the US faculty and have the potential of having a deleterious effect on the quality and quantity of US life science. Still, only somewhat more than 2% of faculty were 65 or older in 1995.

Some data suggest that the lengthening of training is not affecting all segments of the training pool equally. For example, a recent retrospective survey of 192 recipients of the prestigious awards from the Pew Scholars Program in the Biomedical Sciences which identifies promising assistant professors and other research scientists at the beginning of their careers, indicated that their average time to the PhD degree was only 5 years and the duration of their postdoctoral training 3.9 years. The current system has not substantially hampered the rapid progression of these young scientists through training to independent positions, so, at least in this case, it is fulfilling one of its highest priorities: the production of a cadre of truly innovative scientists. But it seems important to do whatever is reasonable to minimize the duration of training while keeping it consistent with the need to prepare young scientists for their careers. It is encouraging that time to degree and age at degree stopped increasing after 1993, but they are still higher than in previous generations of graduates.

EMPLOYMENT PROSPECTS OF YOUNG LIFE SCIENTISTS

The increase in the size of the American postdoctoral population, which has been further increased by the foreign nationals who are training in the United States at both the graduate and postgraduate levels, has led to intense competition for the permanent positions in every sector of the job market, but especially in universities and 4-year colleges. University faculty search committees report hundreds of applications for single positions. Competition among postdoctoral fellows for limited employment opportunities is considered by some to be an ideal way to bring out the best in each person and to select the best people for the jobs. At some critical point, however, competition ceases to bring out the best among aspiring members of the field and becomes a destructive force, breeding conservatism and, at its worst, even dishonesty. When they start new projects, young investigators contribute to an expansion and diversification of the questions being studied in life science. Today, in our experience in the laboratory and on review panels, instead of broadening the fields of inquiry, young investigators are tending to stay within conventional boundaries. If that trend continues, it will ultimately have an adverse effect on the quality of the life sciences.

Our profession must face the fact that current training practices are inexorably leading to increasing problems for the life sciences, not just a crisis of expectation among the young. The issue comes into sharp focus when we take into account the fact that the life-science PhD population problem is going to get worse. The 42% increase in PhD production is a recent phenomenon, and most of the new PhDs have not yet faced the permanent job market, much less begun to compete for grants. Yet the committee's review of future hiring in the life sciences, detailed in chapter 4, provides little likelihood of short-term solutions to the imbalance between PhD production and jobs.

The key to the issue might be in the research and training system now so entrenched. Representative George E. Brown, Jr., the ranking Democrat on the House Committee on Science, has pointed out that with the end of the Cold War, and the slowing of the increase in government investment in research and development, the US science establishment needs to reassess the traditional link between academic research and graduate education (Brown 1997). He argues that the continued linkage means that the number of PhDs produced reflects the availability of academic R&D funding, rather than being related to a set of national goals with respect to the need for science and engineering PhDs. He argues further that we are not analyzing the needs sufficiently and that the result is that production of PhDs can exceed the needs.

This committee's findings support Brown's views on the relationship between research funding and the number of PhDs produced. Life-science research funding has continued to rise in the last 20 years–albeit more slowly than in earlier decades–and PhD output has more than kept pace. Increased research funding means greater demand for workers in laboratories–more graduate students and post-doctoral fellows. But the research-education link also pushes more trained persons into the job market than the available positions in academe, industry, and government can accommodate. This committee's exploration of the nexus between training and the job market has convinced us that the question of national needs is complex and subtle. Although analysis of national needs might not have been sufficient, we note that the problem has defied full solution for 2 decades, because of missing or incomplete evidence, because of the costs of a fuller analysis, and for other reasons–sometimes government rules and procedures themselves. Regardless of the history, we agree with Brown's argument that a reassessment of the nation's linked training and research policies would be useful.

It is plausible that job prospects of young life scientists will diminish further in the coming years unless unforeseen events intervene. The training system, by virtue of its time between graduate-school admission and obtaining of a first permanent position, is slow to respond to changing conditions. It behooves the profession to act in an intelligent and balanced way so that a future crisis will be avoided. If the difficulties of finding appropriate employment become sufficiently widespread, the discontent of postdoctoral fellows might infect undergraduates, who are considering graduate education in life sciences, and result in a decline in high-quality applications. For the future health of the life-science enterprise, we must encourage and retain our most talented aspirants, the people who will always have many attractive options.

In conclusion, the current life-science training enterprise is producing about 2.5 times the number of PhDs needed to fill the jobs that are currently available in academe and when all forms of research-oriented employment are considered, there are still more trainees than there are positions available–and the number of trainees is going up. The recommendations in chapter 6 are designed to ameliorate the stresses in the current situation and to increase the likelihood that we can keep the American life sciences strong and productive.

REFERENCES

Brown, GE. American Physical Society News Online. http://www.aps.org/apsnews/aug97.html

COSEPUP (National Academy of Sciences, Committee on Science, Engineering, and Public Policy). 1995. Reshaping the Graduate Education of Scientists and Engineers. Washington, DC: National Academy Press.

IOM (Institute of Medicine). 1984. The Career Achievements of NIH Predoctoral Trainees and Fellows. Washington, DC: National Academy Press.

McIntosh R. 1994. Funding constraints and population growth: the cell biologist's nightmare. Amer Soc Cell Biol Newsl 17(11):1-5.

NRC (National Research Council). 1995. Research Doctorate Programs in the United States. Washington, DC: National Academy Press.

NSF (National Science Foundation). 1996. Science and Engineering Indicators 1996. NSB 96-21. Washingotn, DC: US Governement Printing Office.

Research!America. 1997. Public Disagrees with Clinton Budget Proposal: Medical Research Funding Should Be Dubled. http://www.nicom.com/~ramerica/newbrief.html.

Stephan PE, Levin SG. 1993. Age and the Nobel Prize Revisited. Scientometrics 28:387-99.

6 CONCLUSIONS AND RECOMMENDATIONS

The committee's study of early research careers in the life sciences revealed a flourishing, productive research enterprise with little unemployment but with a workforce heavily concentrated in "training" positions, such as graduate students and postdoctoral fellows. The occupants of these positions are taking longer to obtain their PhDs; they continue their training after graduate school by assuming postdoctoral positions; their tenure in these postdoctoral positions is lengthening; and when they seek out permanent positions, they face stiff competition–hundreds of applicants for a single post. The net effect of those trends is an ever-growing accumulation of highly trained young scientists in positions that were intended to be transitional. Yet these very people are essential for the accomplishment of the research that has brought so much benefit to the nation and reputation to its life-science endeavor. The committee was faced with an inherent conflict: the system is producing more PhDs than can be absorbed into the permanent workforce, and these trainees are essential to the conduct of research in US universities.

The current situation is the product of a linked education-research system that is in disequilibrium because of features that are intrinsic and structural, that are not confined to the life sciences but have parallels elsewhere in higher education, and that are likely to continue to produce the same outcomes that we have just summarized.

The situation has been building for a long time. In this country, the training of PhDs in science and the performance of scientific research are intimately linked. It has been an article of faith–at least since the 1945 Vannevar Bush report–that both the body of scientific knowledge and the aptitude of young scientists benefit from this linkage. Accordingly, because graduate students play an important role in research projects, the level of graduate enrollments has been strongly influenced by growth in the research enterprise. The arrangement served the nation and the people involved very well during the period of rapid growth in the academic sector that began in the late 1950s. New programs, new departments, and new universities were eager to hire new PhDs (and these new units soon began graduate education programs of their own). By the middle 1970s, however, the growth in the system had begun to slow and it has never regained its earlier rate. Yet the number of new PhDs per year continued to rise (albeit at a much slower rate) while new academic jobs became scarcer. As those two trends continued through the 1970s and the early 1980s, the term of predoctoral study began to lengthen and the proportion of new PhDs who took postdoctoral appointments began to increase, as did the length of time they spent in that status–a sign of the imbalance. To be sure, a substantial increase in hiring in the pharmaceutical and biotechnology industries for a period in the 1980s helped to absorb some of the excess of trained scientists, but that too slowed by the end of the decade. The current situation has been exacerbated by a dramatic 42% increase from 1987 to 1996 in the annual number of PhDs awarded in the life sciences, a substantial proportion of which were awarded to foreign-born candidates. In the same period, the size of the postdoctoral pool grew as well, augmented by an influx of foreign-trained scientists.

Most of the stakeholders in the life-science community are well served by the present arrangements and are likely to be satisfied with how the system is working. The principal exceptions are the senior graduate students and the postdoctoral fellows who are searching for research jobs with career-ladder prospects in academe, industry, or government where they can apply their lengthy training and experience. The search is perhaps most difficult for those who aspire to the university positions toward which their mentors and the academic culture guided them. Although the academic sector is the largest employer of life scientists, the number of openings there and the growth in new positions were being outstripped by the growth in the applicant pool.

Is there any need to intervene, to attempt to redress the imbalance in the system? Some say No–the system is Darwinian, and the competition for occupational survival will bring the fittest to the top. Indeed, the system is designed to winnow out the less competent; not everyone has the talents to become an independent investigator, and it is assumed that some fraction of the graduates will eventually decide to pursue other careers. The system is functioning as it should, and market forces should be allowed to prevail.

This committee takes a different position. We believe that the current rate of production is too high and certainly should not grow higher. The system of training and research that worked so well in times of overall expansion of the enterprise is increasingly deleterious in an era of little growth. The aging of the "young" scientist is disquieting. The system is delaying independence and muffling creativity at perhaps the most productive phase of the individual scientist's life. Finally–and most important–the committee is concerned that an unduly crowded labor market with small chances for success could in the long run drive out the most talented and ambitious aspirants, who will opt for more promising career opportunities in other fields and professions. When the system produces an imbalance like the contemporary one, it is inefficient, wasteful, and dispiriting to its recruits.

For those reasons, the committee believes that there is justification for intervention to adjust the imbalance in the education and training system. At the same time, we recognize the complexity of the system and the diffuse interdependence of its components. In the sections that follow, we report a variety of strategies that the committee has considered for making adjustments, asking of each strategy not only what good purposes it might serve but also what ramifications, especially unwanted consequences, it might have. We have grouped the strategies according to what we believe are desirable goals for making a start on alleviating current difficulties. Overall, our aim is to ensure the continued health of the research enterprise while confronting the disequilibrium that has created a crisis of expectations in the young cohorts who represent the future of life science. We hope that our analysis will focus on the systemic factors that led to the present dilemma and will stimulate widespread discussion in the scientific community about desirable changes.

RESTRAINT OF THE RATE OF GROWTH OF THE NUMBER OF GRADUATE STUDENTS IN THE LIFE SCIENCES

Over the last 2 decades, there has been a substantial growth in the number of life scientists in all categories of impermanent employment[1] owing in no small measure to a sharply increasing number of PhDs being awarded by US universities to both US citizens and foreign nationals, especially in the last decade. This

---

[1] We define the goal of graduate education and postdoctoral training in the life sciences as the preparation of young scientists for careers in independent research in academe, industry, government, or private research environments. We call these "permanent", although it is understood that no employment is guaranteed, to distinguish these positions from the "impermanent" positions, such as postdoctoral fellow and research associate positions held by persons whose career objective is to obtain permanent positions.

growth, which has outstripped the small increases in the number of permanent positions available, has been a major contributor to the swelling of the postdoctoral pool of life scientists. The pool numbers about 20,000, many of whom are marking time until they can move into permanent positions.

Recommendation 1: The committee recommends that the life-science community constrain the rate of growth in the number of graduate students, that is, that there be no further expansion in the size of existing graduate-education programs in the life sciences and no development of new programs, except under rare and special circumstances, such as a program to serve an emerging field or to encourage the education of members of underrepresented minority groups.

The current annual rate of increase in awards of life-science PhDs–5.1% from 1995 to 1996–if allowed to continue, would result in a doubling of the number of such PhDs in just 14 years. Our analysis suggests that that would be deleterious to individuals and the research enterprise. The committee recognizes that the number of PhDs awarded each year might already be too high. Although a return to pre-1988 levels of training might be beneficial, we believe that a concentrated effort to reduce the size of graduate-student populations rapidly would be disruptive to the highly successful research enterprise. The professional structure of life-science research requires the services of graduate students and postdoctoral fellows to conduct the research that is now being funded. A serious reduction in this labor force would impair, delay, or forestall the accomplishment of current and future research.

We caution that it will be necessary to distinguish among fields when making decisions about optimal numbers of graduate students. As shown in chapter 2, almost all the increase in life-science PhD production has been in biomedical fields. Actions taken in one field of the life sciences might be unnecessary in others. It is worth noting, however, that the data shown in figure 3.10 suggest that biomedical and nonbiomedical life-science fields are experiencing similar changes in employment trends, for example, smaller fractions of PhDs finding permanent employment in academe.

The committee acknowledges that its recommendation to constrain further growth will not be easy to implement. Life-science faculties need teaching assistants and research assistants, and limiting the number of entering graduate students will be resisted. But the current rate of growth can no longer be justified, and the premises that have produced it must be reexamined. The committee urges life-science faculties to seek alternatives to these workforce needs (see below in this chapter).

The committee examined several approaches to stabilizing the total number of PhDs produced by life-science departments beyond the first and obvious approach of individual action on the part of graduate programs to constrain growth in the number of graduate students enrolled. As the increases over the last decade, as shown in chapter 2, have been fueled primarily by the increased availability of federal support for research assistants, federal agencies might restrict the numbers of graduate students that they support through the research grant mechanism. If further restrictions were placed by the National Institutes of Health (NIH) on the total amount of salary and tuition support provided for students on research grants well below the current $23,000 cap, it could reduce the attractiveness of research grants as a means of supporting graduate students, although it might also penalize many outstanding programs in private institutions that have high tuitions. Before any action of this sort is adopted, the federal agencies must carefully consider what impact it is likely to have on the university departments and the research efforts being supported.

An alternative approach to restraining the rate of PhD growth is to try to influence career decisions made by prospective graduate students. That could be accomplished, at least in part, by providing accurate and up-to-date information about job prospects for those considering careers in the life sciences. To be sure, the career choices made by students are individual decisions based on a variety of factors, including the attractiveness of alternative career opportunities, the availability of financial support, and a host of personal circumstances. Nevertheless, the most prudent way to reasonably reduce the rate of increase in the number of PhDs awarded annually and perhaps to achieve a gradual reduction in the numbers being trained is to help students to make informed decisions about their career choices. The kinds of information that might be provided and how it might best be compiled are discussed in the next section.

DISSEMINATION OF ACCURATE INFORMATION ON THE CAREER PROSPECTS OF YOUNG LIFE SCIENTISTS

Recommendation 2: The committee recommends that accurate and up-to-date information on career prospects in the life sciences and career outcome information about individual training programs be made widely available to students and faculty. Every life science department receiving federal funding for research or training should be required to provide to its prospective graduate students specific information regarding all predoctoral students enrolled in the graduate program during the preceding 10 years.

Several groups have recognized the need to provide prospective graduate students accurate and up-to-date information on career prospects. As early as 1982, a National Research Council committee studying the employment opportunities for postdoctoral fellows in all fields of science and engineering recommended that the National Science Foundation (NSF) expand its national data-gathering effort to include a survey specifically focused on career decisions of young scientists and engineers. In 1995, a report of the National Academy of Sciences' Committee on Science, Engineering, and Public Policy on graduate education in science and engineering concluded that academic departments should provide employment information and career advice to prospective and current students in a timely manner. Despite those and many other calls for better career information, most life-science students today must rely primarily on the anecdotal reports of their mentors and fellow students.

The earlier recommendations stressed the importance of information for current and prospective graduate students but this committee believes that such data would be equally valuable to faculty, university administrators, and federal policy-makers. In particular, the committee is concerned that the goals discussed here might never be achieved unless the entire life-science community understands fully the implications of the employment trends.

The committee has considered several options to achieve the goal of improved career information. The first is to disseminate widely the data presented in this report. Chapter 3 and the appendixes contain a wealth of information about employment trends over the last 2 decades for young PhDs in the life sciences. Nevertheless, these data have important limitations. First and foremost, because the findings from the Survey of Doctorate Recipients are based on less than 10% of the PhD population, reliable estimates are not available for graduates in a particular discipline, department, or ethnic group.

Thus, although the demonstrated global trends could be useful to policy-makers, they are not especially helpful to faculty advisers and their students who are considering individual career decisions.

A second option would be to expand the sample of recent graduates included in NSF's national survey. Because in recent years this survey has obtained a relatively high response rate (greater than 80%), an expansion of the sample might be expected to yield high returns. The committee regards this step to be valuable but it might not be sufficient to meet all the information needs. For example, reliable data on the early careers of graduates from particular departments would not be available unless a very large sample of recent graduates were selected-and the costs of such a large sample would probably be prohibitive.

A third option that the committee strongly endorses would be to require every department that receives federal funding for research or training to provide current employment information on all predoctoral students enrolled in its program during the preceding 10 years. Such information might include

- The number of trainees and their sex, citizenship, and ethnicity.

- The number of students who left the program before completing their training.

- The length of time from enrollment to degree for each student.

- The current employment situation of each graduate.

One of the major obstacles in implementing a national data collection of such magnitude would be making certain that all federally supported departments provide accurate and comprehensive information that is in a standard format so that comparisons among different departments can be made. Although the difficulty of obtaining reliable information on the current employment situations of graduates from 10 years earlier should not be underestimated, the task is feasible, as demonstrated by the fact that this information has long been a standard requirement for university programs applying for NIH training grants.

A fourth option would be to ask professional societies to assume greater responsibility for compiling and disseminating early-career information. In several science fields (such as chemistry, mathematics, and physics), the professional society conducts a survey of recent doctorate recipients and reports median starting salaries, unemployment rates, and other market indicators. Such a survey would be more difficult in the life sciences because no professional society covers all the disciplines. Nevertheless, professional societies in the life sciences could play active roles in disseminating the information collected by any of the approaches described above. And indeed the committee notes that the Federation of American Societies for Experimental Biology has already published some findings from an analysis similar to that presented in chapter 3 of this report.

IMPROVEMENT OF THE EDUCATIONAL EXPERIENCE OF GRADUATE STUDENTS IN THE LIFE SCIENCES

In addition to its interest in constraining the further growth of PhD output, the committee was concerned about aspects of the current system of supporting graduate training, especially the growth in the fraction of graduate students who are employed as research assistants by the research grants of their mentors. The federal government supports about one-third of all life-science graduate students at any time and about two-thirds at some time in their training, most through salary and tuition provided in the research grants of faculty mentors. That category of student support accounted for the largest percentage of the increase in graduate-student enrollment over the last decade.

There is no clear evidence that career outcomes of persons supported by training grants are superior to those of persons supported by research grants (see discussion in chapter 5). However, the committee, which included members with direct experience with training grants, concluded that training grants are pedagogically superior to research grants and result in a superior educational climate in which students have greater autonomy. First, training grants are pedagogically superior because they provide a mechanism for stringent peer review of the training process itself, something that is not considered in the review of a research project. Second, they improve the educational climate because they minimize the potential conflicts of interest that can arise between trainers and trainees. Although the student-mentor relationship is ordinarily healthy and productive for both partners, it can be distorted by the conditions of the mentor's employment of the student and limit the ability of students to take advantage of opportunities to broaden their education. Third, training grants provide the federal government with information that it needs to evaluate the level of its investment in graduate life-science education with the aim of developing a funding framework for graduate education that contributes to the long-term stability and well-being of the research enterprise.

Recommendation 3: The committee encourages all federal agencies that support life-science education and research to invest in training grants and individual graduate fellowships as preferable to research grants to support PhD education. Agencies that lack such programs should look for ways to start them, and agencies that already have them should seek ways to sustain and in some instances expand them.

This recommendation should not be pursued at the expense of scientific and geographic diversity. Rather, we encourage the establishment of small, focused training-grant programs for universities that have groups of highly productive faculty in important specialized fields, but might not have the number of faculty needed for more traditional, broad-based training grants.

It is true that the current regulations governing NIH training grants bring universities some financial disadvantages because of restricted overhead recovery. Furthermore, NIH training grants cannot support foreigners on student visas, and so this recommendation places at disadvantage programs that depend on foreign students for research or teaching. These disadvantages are outweighed, in the committee's view, by the salutary effect that the training-grant peer-review process brings to the members of a department faculty, leading them to examine and reflect on how, as an entity, they are providing for the education and training of their graduate students.

Our endorsement of training grants and fellowship is not intended to result in the training of more PhDs, which we argue would be entirely inappropriate. Rather, any growth in the numbers of trainees supported through an expansion of training grants should come at the expense of the numbers of trainees supported on research grants. Thus, the implementation of this recommendation should produce no increase in the numbers of students but only a change in the mechanism by which their training is supported by federal funds. It would be best if principal investigators voluntarily reduced the number of students they support on their research grants as support via training grants grew. However, NIH, the largest provider of both training grants and research grants, and other agencies would be required to manage the numbers supported by research grants to achieve the committee's goal of constraining further growth.

The committee is also concerned that the length of time spent in training has become too long, at a median of 8 years of elapsed time from first enrollment to PhD in all the life sciences though field differences exist. We believe that the time should be about 5-6 years. However, an immediate effort to shorten the time to degree would increase the number of PhDs produced. Efforts to shorten the time to degree should be undertaken when the effort to restrain growth in the number of PhDs has shown positive effects.

ENHANCEMENT OF OPPORTUNITIES FOR INDEPENDENCE OF POSTDOCTORAL FELLOWS

While the length of graduate training has been increasing, so too have the extent and duration of postdoctoral training. Prolonged tenure as a postdoctoral fellow provides a person with valuable research experience, but it carries some real costs. In most cases, fellows are not independent of their mentors so they can not pursue their own research. We recognize the many good reasons for prolonged tenure as a postdoctoral fellow but we believe that tenures longer than 5 years are not in the best interest of either the individual fellow or the scientific enterprise.

Unfortunately the committee did not identify a way to rapidly achieve a reduction in the tenure of postdoctoral fellows. The lengthening of the postdoctoral period seems to be due largely to the highly competitive job market for permanent positions in academe and industry; the situation will change only if there is an increase in the number of new positions or a decrease in the candidates for them.

Recommendation 4: Because of its concern for optimizing the creativity of young scientists and broadening the variety of scientific problems under study in the life sciences the committee recommends that public and private funding agencies establish "career-transition" grants for senior postdoctoral fellows. The intent is to identify the highest-quality scientists while they are still postdoctoral fellows and give them financial independence to begin new scientific projects of their own design in anticipation of their obtaining fully independent positions.

The recommendation is based on the experience of the Lucille P. Markey Charitable Trust's Scholars in Biomedical Sciences Program, which until recently supported 16 postdoctoral fellows per year for 2 years of additional postdoctoral work and 5 years as faculty members. Although the program was very small, it identified excellent candidates relatively early in their careers and gave them financial and intellectual independence. Not surprisingly, the Markey scholars were very successful in obtaining permanent tenure-track positions in academe. Since the termination of the Markey program, the Burroughs Wellcome Fund has established a comparable program for life scientists. A program administered by the US Department of Agriculture provides postdoctoral fellows the opportunity to apply for research grants and perform independent research.

We propose grants of 4-5 years in duration that would provide senior postdoctoral fellows (those with more than 2 years of postdoctoral experience) salary commensurate with their experience and a modest supply budget. Successful proposals would define an innovative research project that was distinct from the work going on in the current mentor's laboratory. A mentor would provide laboratory space and would acknowledge in the applications that the project was the intellectual property of the applicant and would leave the laboratory when the applicant did.

The committee recommends a goal of 200 federal grants awarded annually, representing about 1% of the postdoctoral pool. That number of people supported would be quite small but the program might provide an important opportunity for the most promising postdoctoral fellows and serve as both example and incentive to many more. We make this recommendation with the knowledge that it is possible that the money for a new federal grant program probably would come from existing federal funds. In our view, the benefits of increased intellectual independence and improved motivation of talented midcareer post-doctoral fellows justify such a reallocation of funds. Private funders might establish new programs or enlarge existing programs that support career-transition grants.

The career transition grant would differ from existing federal research grants in several important ways. First, permission to apply for traditional grants is usually restricted by institutions to principal investigators who have some form of faculty status, whereas these new grants would go to postdoctoral fellows. Second, the career-transition grants would be modest in scale and would not provide salary support for other laboratory personnel or trainees. Finally, the grants would be transferable to new host institutions once the applicants obtained positions and would terminate on receipt of faculty awards. The success of this recommendation depends on a willingness of training institutions to accept grants to persons who do not have faculty status at the time of application.

The benefit of career-transition grants to individual young scientists is obvious: increased independence means increased opportunity to pursue novel ideas and to make progress in work that can establish a career, opening opportunities for future independent employment. Substantial benefits would also be realized by the scientific enterprise as a result of this stimulation of research energy and the increased diversity in the scientific ideas being pursued. Less obvious but no less important is the benefit that would accrue to the mentors. The presence of more experienced scientists in the host laboratories, although not directly contributing to the productivity of the mentors' work, will contribute to the intellectual climate of the laboratories.

ALTERNATIVE PATHS TO CAREERS IN THE LIFE SCIENCES

As traditional research positions in academe, industry, and government have become more difficult to obtain, positions in "alternative careers"–such as law, finance, journalism, teaching, and public policy–have been suggested as opportunities for PhDs in the life sciences.

The idea of highly trained PhD scientists investing their talents in nontraditional careers seems at first glance attractive. Scientists have analytic skills and a work ethic to bring to any position, and the placement of highly trained scientists in diverse jobs in the workforce would lead to an increase in general science literacy. As the committee's review of alternative opportunities (chapter 4) concludes, however, most of the possibilities are less available or less attractive than they might at first glance appear. Many "alternative" careers are also heavily populated, and competition for good positions is stiff. Others require special preparation or certification or offer unattractive compensation, and none makes full use of the PhD's hard-won life-science research skills. The committee believes that the idea of alternative careers should not be oversold to PhD candidates.

Conclusions and Recommendations

The interest in alternative careers for PhD scientists has inevitably raised the question of whether preparation for the degree should be changed from its current narrow focus on training for the conduct of scientific research to embrace a broader variety of educational goals that would connect to alternative career paths. The committee has discussed that question extensively.

Recommendation 5: The committee recommends that the PhD degree remain a research-intensive degree, with the current primary purpose of training future independent scientists.

We have several reasons for that recommendation. First, a steady supply of new, highly trained investigative talent is essential for maintaining the growth and vigor of life-science research and for exploiting the opportunities of future discoveries. Second, the majority of people so trained are using their skills and abilities in life-science positions. Third, we have not been able to identify a substantial number of unfilled opportunities in alternative careers.

At the same time, the committee recognizes that not all students who begin graduate school intending to pursue research careers maintain that desire as they progress through training. Graduate programs should expand their efforts to help students to learn about the diversity of career opportunities open to them, and university departments should examine possible alternatives to the research PhD, for example, rigorous master's-degree programs in applied fields of the life sciences.

The master's degree might be a more appropriate end point for students who determine early enough in their training that PhD training is not necessary for the career goals that they have selected. There has been a decline in the number of master's-degree programs in the life sciences and with it a growing perception that the master's degree has become a consolation prize for those who do not complete a PhD program. Those changes effectively limit the number of choices for college graduates who are interested in a career in the life sciences, although not necessarily careers in directing laboratories conducting fundamental research.

Recommendation 6: The committee recommends that universities work to identify specific fields of the life sciences for which master's-degreee training is more appropriate, more efficient, and less expensive than PhD training and that focused master's-degree programs be established in those fields.

A reinvigoration of the master's degree will require that new programs be intimately tied to the opportunities in the labor market. For example, a life scientist who is interested in a K-12 or 2-year-college teaching career would benefit from formal and focused master's-degree programs that do not require long periods of research-intensive graduate and postdoctoral training. In chapter 4, we report that life-science PhDs have not been prone to take positions as precollege teachers. Certainly, there is a need for persons with life-science knowledge to enter teaching careers. Intensive efforts are under way to change the nature and extent of science education in our schools. Those efforts, based on the National Science Education Standards and similar reform documents, emphasize teaching science as inquiry rather than as word associations. None of this will be possible without a structural change in the profession of precollege teaching and a large cadre of people who both understand science and the nature of science as inquiry and have been trained as lead teachers and science-resource specialists. Focused and intensive master's-degree programs would be not only more appropriate but also preferable to the PhD for this type of employment.

Interdisciplinary master's-degree programs

might combine advanced life-science training with studies in nonscientific fields–such as management, public affairs, and engineering–that would prepare candidates for positions in government and industry. A vigorous master's-degree program that produces highly skilled laboratory technicians for industry, government, and academe could potentially contribute to righting the imbalance between PhD training and the labor market. When the committee recommended constraint in further growth in training in recommendation 1, it was fully aware that graduate students are needed in the labor-intensive life-science research enterprise and to teach undergraduates. One way to resolve this dilemma is to effect a modest shift toward a more permanent laboratory workforce by replacing some fraction of the existing training positions with permanent employees, such as MSc-level technicians and PhD-level research associates.

A system of that kind, with less reliance on trainees to conduct research, has been in operation in Europe for many years. Nevertheless, there is likely to be strong resistance to such a change in the US scientific community. Permanent employees would require better compensation in the form of salary and benefits than graduate students and postdoctoral fellows and could not be expected to work the long hours of most trainees. As a consequence, a shift to a more permanent workforce would probably result in some reduction in productivity and cost effectiveness. Furthermore many US scientists are of the opinion that the creativity of US science comes from the young and inquiring minds of young trainees. Despite those reservations, the committee believes that a broader discussion of this option within the life-sciences community is warranted.

THE IMPACT OF FOREIGN NATIONALS

This report has documented that much of the recent increase in the number of life-science PhDs granted by US universities are foreign nationals, not US citizens–in some years, as much as one-fourth of the degrees awarded. The number of foreign nationals reflects the international nature of modern science and the central place that the United States plays in this international arena. Furthermore, foreign nationals have traditionally contributed to the excellence of US science, as suggested by the fact that of the 732 members of the National Academy of Sciences who are life scientists, 21.2% are foreign-born and 12.4% obtained their PhD training abroad. Foreign nationals' important contributions to US scientific leadership is reflected in their inclusion as "outstanding authors" in life sciences (26.4%). Foreign students and fellows are welcome participants in the research enterprise, provided that they are of high quality and competitive with American applicants.

Although the reasons for the increase in degrees awarded to foreign nationals are not altogether clear, the committee understands that it is a phenomenon essentially controlled by life-science departments themselves, inasmuch as immigration law virtually delegates visa decisions to universities. Departments and universities make their own admission and funding decisions and issue documents to those they admit, which nearly always results in the US government's issuing student visas (subject to checks for fraud and funding availability). The freedom given to US universities to determine how many foreign students they will admit carries responsibilities. If misused, it could vitiate the committee's recommendation to provide up-to-date and full career information to prospective applicants for graduate education in the life sciences. That information might have a powerful effect on US citizens but it is highly unlikely to have a similar effect on students from low-wage economies with poor educational or research opportunities. Even the low stipends paid to graduate students enable a higher standard of living for such applicants; and the prospect of a job or postdoctoral position and a permanent visa at the completion of graduate study is a powerful incentive for citizens of many countries.

We believe it would be unwise to place arbitrary limits on the number of visas issued for foreign students. But we do not believe that US institutions should continue to enroll unlimited numbers of foreign nationals. As decisions are made on ways to constrain further growth, the measures adopted should apply equally to all students regardless of nationality.

Recommendation 7: If, as we hope, implementation of our recommendations results in constraining further growth in PhDs awarded in the life sciences, we urge our colleagues on graduate admissions committees to resist the temptation to respond by simply increasing the number of foreign applicants admitted.

Postdoctoral fellows are also recruited from abroad. At present, half the roughly 20,000 postdoctoral fellows in the United States are foreign nationals, many of whom entered the country with PhDs awarded elsewhere. These scientists constitute an important part of the research labor force, as well as of the pool of applicants for permanent jobs in academe, industry, and government. In this instance again, we urge our colleagues to give equal opportunity to US citizens and foreigners and to refrain from hiring foreign nationals to fill the places of US scientists.

RESPONSIBILITY FOR EFFECTING CHANGE

This report has documented several dramatic changes in career trends in the life sciences over the last several decades. The rapid growth in the academic scientific establishment in the 1960s and the early 1970s set in place a training infrastructure that was built on the premise that there would be continued growth. When the inevitable slowdown in resources to support that growth occurred, it was not accompanied by a commensurate adjustment in the rate of training. The impact of the imbalance between the number of aspirants and the research opportunities is now being felt by a generation of scientists trained in the last 10 years who are finding it increasingly difficult to find permanent positions in which their hard-accumulated skills in research can be used. Unless steps are taken to put the system more in balance, the difference between students' expectations and the reality of the employment market will only widen and the workforce will become more disaffected. Such an occurrence would damage the life-science research enterprise and all the participants in it.

The training of life scientists is a highly decentralized activity. Notwithstanding the heavy dependence on federal funds, the most important decisions affecting the rate of production of life scientists are made locally by the universities and their faculties. The numbers and qualifications of students admitted to graduate study, the allocation of institutional funds for their tuition and stipends (which account for half or more of the total expenditures for graduate-student support), the requirements for the degree—all are local decisions. As a consequence, a large portion of the responsibility for implementing our recommendations falls on the shoulders of established investigators, their departments and universities, professional scientific organizations, and students themselves. Students must take the responsibility of making informed decisions about graduate study, but they must be provided accurate career information on which to base their decisions. Individual faculty members must be willing to set aside their short-term self-interest in maintaining the high level of staffing of their laboratories for the sake of the long-term stability and well-being of the scientific workforce. Directors of graduate programs must be willing to examine the future workforce needs of the scientific fields in which they train, not just the current needs of their individual departments for research and teaching assistants.

The recommendations in this report are offered as first steps to improve the overall quality of training and career prospects of future life scientists. We hope that the information in this report will be used to begin discussions within the life-science community on the best ways to prepare future scientists for exciting careers in the profession and to protect the vitality of the life-science research enterprise.

An Alternative Perspective on Recommendation 3
Henry W. Riecken

Let me begin by stressing that I dissent from the unqualified endorsement and recommended expansion of training grants in chapters 5 and 6 and *not* from the overall study findings, which I strongly support. The compelling evidence presented in chapters 2 and 3 and appendixes, together with the confirming testimony at the public meeting and experiences of individual committee members, led us to the unanimous conclusion that the current level of PhD production now exceeds the current availability of jobs in academia, government, and industry where new life-science PhDs can independently use their training. We also unanimously agreed that further growth in graduate training in the life sciences must be curtailed and that there should be no further expansion of graduate educational programs except "under rare and special circumstances".

The committee had a much more difficult time, however, in deciding how best to achieve the recommended goal of stabilizing graduate enrollments. The difficulty derives chiefly from the complex interdependence of research and training, as described in chapter 6. While some of the committee's recommended actions–in particular, the broad dissemination of information pertinent to career prospects–will be useful in addressing this goal, I strongly disagree with the recommendation to increase training-grant support. In my view, this recommendation is unsupported, outside the study charge, and inconsistent with the committee's overall study findings. My specific objections to this recommendation are as follows:

(1) Recommending that federal agencies expand training-grant programs conflicts with the committee's desire to stabilize graduate enrollments. While the report states that "the expansion of training grants should come at the expense of the numbers of trainees supported on research grants", the committee offers no guidance to the federal agencies on *how* to reduce the number of federally supported research assistants. At the second meeting of the committee, an NIH official told us that the agency had no control over the total number of students supported on research grants since they are essentially employees hired by the universities and principal investigators. Absent effective control on the number of students supported on federal research grants, the recommended expansion of training grants would increase the availability of federal support for graduate education and likely lead to an increase in graduate enrollments–precisely what the committee wishes to halt.

(2) The recommendation to reduce support for research assistantships (while increasing training grants) also conflicts with the committee's expressed opinion that it would be unwise to impose limitations on the admission of foreign nationals to graduate study in US universities. Since foreign students are not eligible for training-grant support, the total amount of support available to them would be diminished by the proposed substitution of traineeships for research assistantships–thereby limiting their access to training in the United States.

(3) The committee was *not asked* to evaluate the quality of predoctoral education or the relative merits of alternative mechanisms for support of graduate training. In fact, at the outset NIH officials made it clear that this study should not duplicate the efforts of the National Research Council Committee on National Needs for Biomedical and Behavioral Research Personnel, which was established at the request of Congress and explicitly charged with recommending the level of training-grant support provided by NIH. The recommendation to expand training-grant support clearly intrudes on this other Research

Council committee's work.

(4) The committee did not investigate systematically and carefully the advantages and disadvantages of alternative mechanisms of predoctoral support. The only factual evidence pertinent to this issue (presented in chapter 5) comes from a 1984 Research Council study, *The Career Achievements of NIH Predoctoral Trainees and Fellows*. This study explicitly stated that "it cannot be determined whether [trainees'] superior records of achievement may be attributed to the selection process, the training they received, or a combination of these and other factors." Thus, any conclusion drawn from this study that training grants are a more effective training mechanism than research grants is unfounded.

(5) The report's stated preference for training grants over research grants is not based on hard evidence of superiority, but rather on the opinions of individual committee members "with direct experience with training grants". Since the study charge does not encompass an evaluation of alternative mechanisms for graduate student support, it is not surprising that a majority of the committee do not have such "direct experience". They are therefore not in a position to make independent judgments about the relative merits of these two training mechanisms and were not appointed with this task in mind.

(6) The advantages and disadvantages of alternative support mechanisms were never fully discussed by the committee. Had the study called for a comparison of alternative mechanisms for predoctoral support, a much more detailed analysis would have been required, including an examination of the cost implications for different institutions and federal sponsors. (NIH training grants do not pay full indirect costs, while research grants do; and training grants also limit trainees' tuition reimbursement to the university.)

(7) The proposal to substitute traineeships for research assistantships presents a particular problem for institutions that do not have training grants, yet have faculty members who are successful in obtaining NIH research awards. These investigators would be unable to make the recommended substitution, yet the quality of their research can be assumed to be as good as the research funded at universities that do have training grants.

(8) From the perspective of federal policymakers, the recommendation to increase training grant support may appear nonsensical–especially in light of the overwhelming evidence that universities are already training too many PhDs for the research positions available. Why should Congress appropriate more funds for training grants when there is already an overabundance of trained life scientists?

I want to emphasize that I have these reservations about the training-grant recommendation because of the totally inadequate evidential basis for the recommendation and because of the consequences it would have–*not* because I hold strong views on the intrinsic merits of either training grants or research assistantships. For several years, I chaired the aforementioned Committee on National Needs for Biomedical and Behavioral Research Personnel, which recommended annually to Congress the number of training-grant positions to be supported under the National Research Service Awards Act. Earlier, I served as associate director of the National Science Foundation with particular responsibility for the education and training of scientists (in all scientific disciplines). These experiences have made me keenly aware of the difficulty of making a valid comparison between alternative support mechanisms, as well as the multiple difficulties of implementing the changes recommended in this report. Without considerably more evidence on the relative merits of alternative mechanisms for supporting graduate students, a recommendation to increase training grants and substitute these positions for research assistantships is unwarranted–and detracts from what I consider to be an otherwise scholarly and objective analysis.

Appendix A

BIOGRAPHIC INFORMATION

Shirley Tilghman (Chair) is the Howard A. Prior Professor of the Life Sciences at Princeton University and an investigator of the Howard Hughes Medical Institute. She is a molecular geneticist whose work focuses on the regulation of genes during development. She is a member of the Royal Society of London, the National Academy of Sciences, and the Institute of Medicine.

Helen S. Astin is a psychologist, professor of higher education, and associate director of the Higher Education Research Institute at the University of California at Los Angeles. Her research and writings have focused on the education and career development of women and on faculty careers, productivity, and rewards.

William Brinkley is Distinguished Service Professor of Cell Biology, vice president for graduate sciences, and dean of the graduate school of biomedical sciences, Baylor College of Medicine. His research involves studies of mitosis and genome instability in eukaryotic cells. He is interested in PhD education in academic health centers and was the founder of the Association of American Medical Colleges Graduate Research Education and Training Committee whice explores issues also dealt with in this report.

Mary Dell Chilton is Distinguished Science Fellow at Ciba-Geigy Biotechnology, where she continues research on the molecular biology of plant genes. She is a member of the National Academy of Sciences.

Michael P. Cummings was, at the beginning of this study, a postgraduate research plant geneticist in the Department of Botany and Plant Sciences, University of California, Riverside. He is now at the Center for Comparative Molecular Biology and Evolution, Marine Biological Laboratory. His research focuses on empirical and computer-based investigations in molecular evolution, population genetics, and systematics.

Ronald G. Ehrenberg is vice president for academic programs, planning, and budgeting at Cornell University. A member of the Cornell faculty for 21 years, he is the Irving M. Ives Professor of Industrial and Labor Relations and Economics and the author or co-author of over 100 papers and books. He was the editor of *Research in Labor Economics*, and is a co-editor of the *Journal of Human Resources*. He is also a research associate at the National Bureau of Economic Research and a member of the Executive Committee of the American Economic Association. Much of his recent research has focused on higher-education issues. He regularly taught a popular course titled "Economic Analysis of the University".

Mary Frank Fox is professor of sociology, School of History, Technology, and Society, Georgia Institute of Technology. Her research focuses on women and men in scientific and academic organizations and occupations; her current work is a study of gender and doctoral education in five science and engineering fields. Her publications, appearing in over 30 scholarly journals and collections, include analyses of salary, publication productivity, and educational and career patterns among scientists. She is associate editor of *Sex Roles*, past associate editor of *Gender & Society*, and chair of the Editorial Board of the international *Handbook of Science and Technology Studies*.

Kevin Glenn is a fellow in cardiovascular diseases research at Searle. He has served on previous National Research Council committees involved with PhD issues.

Pamela J. Green is associate professor, Michigan State University/Department of Energy Plant Research Laboratory and Department of Biochemistry at Michigan State. Her research focuses on the control of mRNA stability and ribonuclease regulation and function in higher plants. She is past cochair of the North American Arabidopsis Steering Committee and is a member of the Board of Directors of the International Society for Plant Molecular Biology. She has organized "Choices Day" at the Plant Research Laboratory and has contributed to workshops at American Society of Plant Physiologists meetings to inform students about the spectrum of careers in science.

Sherrie L. Hans was a graduate student in the Department of Biochemistry at the University of California, San Francisco until the summer of 1996, when she received her PhD. She was supported by a National Science Foundation graduate fellowship during the first 3 years of her graduate career. Currently, she is a program officer for biomedical research policy at the Pew Charitable Trusts in Philadelphia.

Bruce R. Levin is professor of biology and director of the Graduate Program in Population Biology, Ecology and Evolution at Emory University. Dr. Levin's current research includes theoretical and experimental studies of the population biology and evolution of bacteria and infectious disease. He received his PhD in Genetics from the University of Michigan in 1967. He has taught at Brown University and the University of Massachusetts.

Arthur Kelman is a University Distinguished Scholar in the Department of Plant Pathology, North Carolina State University, and Emeritus Wisconsin Alumni Research Foundation Professor of Plant Pathology and Bacteriology, University of Wisconsin-Madison. His research has been in the area of mechanisms of pathogenesis of bacterial plant pathogens and the nature of disease resistance in plants. He has served as chairman of the Board on Basic Biology, on a number of other committees of the National Research Council, and as chief scientist of the National Research Initiative Competitive Research Grants Program of the US Department of Agriculture. He is a member of the National Academy of Sciences and the American Academy of the Arts and Sciences.

J. Richard McIntosh is professor of cell biology at the University of Colorado, Boulder and a research professor of the American Cancer Society. His principal research interest is the mechanisms by which cells organize and segregate their chromosomes in preparation for cell division. He is also principal investigator of the Laboratory for Three-Dimensional Fine Structure, a national research resource that is developing new technologies for the study of cellular architecture. He has taught cell biology at the graduate and undergraduate levels.

Henry W. Riecken is the Boyer Professor emeritus of Behavioral Sciences at the School of Medicine of the University of Pennsylvania. He is a psychologist who formerly headed the Divisions of Scientific Personnel and Education at the National Science Foundation. He was Chairman of the National Research Council Committee on National Needs for Biomedical and Behavioral Research Personnel. He is a founding member of the Institute of Medicine

Paula E. Stephan is associate dean and professor of economics, School of Policy Studies, Georgia State University. She is a labor economist by training and her recent research focuses on the economics of science and innovation. She has also studied the relationship of age, career stage, and birth origin to productivity. She is the author of over 50 books and papers. She has served as a consultant to a number of organizations and as a visiting scholar at the Wissenschaftszentrum Berlin für Sozialforschung, Berlin, Germany.

Appendix B

PARTICIPANTS IN PUBLIC MEETING

The committee sponsored a public meeting on April 13, 1996, to hear the views of the life-science community on the issues included in the committee's charge. Listed below are the names of speakers at the public meeting and the names of those who attended the meeting or submitted statements for the benefit of the committee.

SPEAKERS

Robyn Angliss, National Marine Fisheries Service
Eliene Augenbaum, Association of Science Professionals
Finley Austin, Burroughs Wellcome Fund
Kevin Aylsworth, Senator Hatfield's Office
John Beneditt, *AAAS/SCIENCE Next Wave*
Carol Brewer, University of Montana
Malcolm Campbell, Davidson College
Rita Colwell, University of Maryland Biotechnology Institute
Glenn Crosby, Washington State University
Caren Helbing, University of Calgary, Canada
Brian Hyps, American Society of Plant Physiologists
Gene A. Nelson, Microsoft Corporation
David Olson, University of California, San Francisco
Erika C. Shugart, University of Virginia
Sam Silverstein, Federation of American Societies for Experimental Biology
Abigail Stack, Food and Drug Administration
Michael Teitelbaum, Sloan Foundation
Robert Tombes, Virginia Commonwealth University

ATTENDED OR SUBMITTED PAPERS

Josephine C. Adams, University College London
Janet van Adelsberg, Columbia University
Stan Amons, Association of American Medical Colleges
Michael Battalora, National Institute of Environmental Health Sciences
Scott D. Blystone, Washington University
David B. Bregman, Yale School of Medicine
Sheryl K. Brining, National Institutes of Health
Shawn Burgess, Massachusetts Institute of Technology
David G. Capco, Arizona State University
Ida Chow, American University
Stan Cohn, DePaul University
David R. Cool, National Institutes of Health
Jaleh Daie, University of Wisconsin
Jerry Dodgson, Michigan State University
Diane Epperson, National Institutes of Health
Evan Ferguson, Sigma Xi
Michael Fordis, National Institutes of Health
Catherine Gaddy, Council on Scientific Personnel
Howard Garrison, Federation of American Societies for Experimental Biology
Ursula Goodenough, Washington University
Jay A. Haron, Knight-Ridder Information
Joanne Hazlett, National Science Foundation
Philip M. Hemken, Iowa State University
Julie R. Hens, University of Maryland
Milton Hernandez, National Institutes of Health
Marc Horowitz, National Institutes of Health
Elizabeth Jansen, University of Minnesota
Naomi Kaminsky, American Pharmaceutical Association
Doug Kellogg, University of California, Santa Cruz
Eero Lehtonen, University of Helsinki
John Lowe, University of Michigan Medical School
R. Joel Lowy, AFRRI/Department of Defense
Anthony C. Madu, Virginia Union University
Michael S. Marks, University of Pennsylvania
Charles Matsuda, University of Hawaii
Bert Menco, Northwestern University
Katsumi Mochitate, National Institute of Environmental Health Sciences
Randall T. Moon, University of Washington
Alan Munn, Switzerland
Richard Murphy, Neurological Institute, Canada

Lynne A. Opperman, University of Virginia
Christine M. Pauken, Food and Drug Administration
Michael Powell, National Institutes of Health
Linda Pullan
Janet Ross, Proceedings of the National Academy of Sciences
Charles Selden, National Institutes of Health
Heidi Sofia, National Institutes of Health
Robert Stack, University of Michigan
W. Steven Ward, New Jersey Medical School
Tracy Ware, University of California, San Francisco
Ora A. Weisz, University of Pittsburgh
Cheryl Wellington, University of Calgary
Marianne Wessling-Resnick, Harvard School of Public Health
Lawrence Wiseman, College of William and Mary
Joyce Woodford, National Institutes of Health
Marie Wooten, Auburn University

Appendix C

SOURCES OF DATA

NATIONAL RESEARCH COUNCIL

SURVEY OF EARNED DOCTORATES

The Survey of Earned Doctorates (SED) is conducted annually by the National Research Council and is a census of the research doctorates awarded at US universities during the academic year, from July 1 of one year to June 30 of the following year. The self-report response rate from the PhD recipients is about 95%, and information on the remaining 5% of the doctorates is obtained from commencement programs and institutional sources. The survey gathers information on all fields that award research and applied-research doctorates, except professional degrees such as the MD, DDS, OD, DVM, and JD. It gathers data on a field-specific basis, and includes information on ethnic background, sex, postsecondary education, time to PhD degree from the baccalaureate degree, financial support during graduate studies, and postdoctoral plans. The data from the survey become part of the Doctorate Records File (DRF), a virtually complete database on doctorate recipients from 1920 to the present. The data in this file can be manipulated in different ways to obtain the characteristics of graduates by nearly 20 broad fields or several hundred fine fields with regard to their institution, their graduate program, and their plans. The data in the DRF are kept on an individual basis and are linked to other files, such as the file for the Survey of Doctorate Recipients (see below) and the National Institutes of Health grants files.

In the life-science fields included in this report, 7,696 doctorates were added to the DRF in 1996. The field specialties in the life sciences include the agricultural and biomedical sciences and a portion of the health sciences as broad fields, and these are divided into 67 fine-field specialties.

Data Considerations

The information in the DRF is complete and reliable for most data points. However, in the case of the data on sources of support during graduate school, students are not always aware of their sources or the type of support, and for postgraduate plans, the survey questionnaire might be complete at a time before a definite commitment or reflect a hope of a particular type of postdoctoral position.

SURVEY OF DOCTORATE RECIPIENTS

The Survey of Doctorate Recipients (SDR) is a biennial longitudinal survey, dating to 1973, of research doctorate-holders working in the United States. The sample for each survey period is adjusted by the addition of persons from the most recent 2-year cohort in the DRF and the dropping of persons who have retired or have reached the age limit of the survey. Before 1991, the population of the survey included a broader range of people, such as holders of US-earned doctorates in humanities, education, and professional fields who were working in science and engineering (S&E), holders of foreign-earned doctorates who were working in S&E in the United States, and a 42-year period of PhD cohorts. The SDR was restructured in 1991 to include only persons under the age of 76 years who hold doctorates in S&E from US universities, and the sample was reduced by 55% to provide resources to increase the response rate.

The survey questionnaire is sent in the spring to each person in the sample. In 1995, the sample numbered 49,829. The people in the sample are asked a series of demographic and employment questions. The response rate for the survey in 1995 was about 85% after second-wave mailings and telephone interviews; this was about a 30% increase in the response rate over 1989. Although the reduction of the sample reduced the overall number of responses from 1989 to 1995, it is believed that the increased response rate improves the quality of the data. However, the change in the survey produced a potential disjunction between data collected before 1991 and those collected since.

The sample is stratified across three variables: field of degree, sex, and a combination variable that includes degree field, sex, handicap status, ethnic group, and nationality of birth. The results of the survey are statistically analyzed to translate the data into weighted numbers for the entire population. From the weighted results, the doctorate workforce in S&E can be analyzed across different dimensions by looking at different demographic and employment characteristics and by taking different cohorts. That provides for both longitudinal and time-series analyses. However, in the analysis, one must take into consideration the change in sampling frame, the increased response rate in 1991, and the fact that some cells in an analysis could contain very few actual responses, in that the sample is only about 8% of the S&E workforce.

Data available from the SDR up to 1991 are field of doctorate and employment, sector of employment, geographic location, primary work activity, federal support, tenure status, salary data, and ethnic data. However, the 1991 SDR was administered in the fall, not the spring; some data points are not directly comparable with those from other survey years. The 1993 questionnaire incorporated substantial changes from earlier ones. In particular, the questionnaire before 1993 asked for data only as of a specific time, but the 1993 questionnaire asked for some retrospective employment information. There was also a change in the field employment questions, with much broader definitions of job categories, such as "biological scientist", as opposed to, for example, "ecologist" in the earlier surveys. As a result, the number of people in postdoctoral positions might have been slightly overestimated. In 1995, additional questions concerning detailed retrospective descriptions of the time spent in postdoctoral training were added.

Data Considerations

The SDR is a sample survey of about 8% of PhD awardees, and the number of responses might be low in some cases. A weighting formula is used to adjust the sample to the complete population. For example, a weighted response of 39 unemployed life scientists from the 26 high-quality institutions in 1995 corresponds to five responses; the 20 people working outside S&E in the same population is based on three responses. In the experience of the National Research Council's Office of Scientific and Engineering Personnel who have worked with these data for many years, a response of 10 or more provides a good estimate for a category. Although the sample is small and the analyses must be used with care, the sampling and weighting methods have been carefully developed to provide the most statistically valid results possible.

NATIONAL SCIENCE FOUNDATION SURVEY OF GRADUATE STUDENTS AND POSTDOCTORATES IN SCIENCE AND ENGINEERING

The National Science Foundation (NSF) conducts various surveys and data-collecting procedures as part of its responsibility in monitoring the state of science and engineering development in the United States. The survey that pertains most closely to graduate and postdoctoral training is the annual Survey of Graduate Students and Postdoctorates in Science and Engineering.

This survey is designed to provide a comprehensive picture of the training of future scientists and engineers in US graduate schools and is used to assess future supply and demand. Graduate students counted in the survey are enrolled for credit in science and engineering master's-degree and PhD programs in the fall term of the survey year, and MD, DO, DVM, and DDS candidates are reported only if they will also receive a PhD. The survey also includes information on postdoctoral appointees and other nonfaculty researchers in academic departments and programs.

The survey is distributed to departments through an institutional coordinator and information is provided on students that are associated with departments. Nearly 10,400 graduate departments at 730 institutions are surveyed. Students in interdisciplinary or interinstitutional programs are reported only by their primary department. Therefore, information about individual programs could be distributed across departments, and data would be aggregated for departments with multiple degree programs.

The following types of information are requested:

- Number of full-time graduate students separated by type of financial support, source of support, and sex, and number of first-year students (no distinction is made between MS and PhD students.
  - Number of part-time students and their sex.
  - Ethnicity of full-time and part-time students who are US citizens.
  - Number of full-time and part-time foreign students.
- Number of postdoctoral and nonfaculty research positions in the department, with type of support for the positions, whether US citizen or foreign, and the sex of the person in each position.

The NSF requests that the survey form be returned by January 31 for data on the previous fall enrollments. The data are reported in a series of reports, many of which are available online through the Internet, on the different aspects of education by institution and field within the institution. However, data tapes will provide more detailed information on separate departments.

Data in table E.3, and figures 2.3 and 2.6 are taken from this NSF survey and are not directly comparable with other data, from the SED and SDR, used throughout the report. The NSF survey counts only persons at academic institutional whereas the SDR counts PhDs in all work environments. Furthermore, NSF definitions of fields differ somewhat from those used in this report (Appendix D). Those differences are not important when addressing questions about graduate students, because students are at academic institutions where NSF performs its survey. However, large differences in the count of postdoctoral fellows can exist between the NSF survey and the SDR. We have used the NSF count of postdoctoral fellows at academic institutions as a starting point because NSF counts both US citizens and foreign nationals, whereas the SDR excludes foreign nationals who have not received their PhD in this country. We have then estimated the number of postdoctoral fellows who might be in government, industry, and other nonacademic laboratories to obtain an estimate of the overall number of postdoctoral fellows in the United States.

Data Considerations

The quality of the survey data depends on the knowledge of the persons at the department level who complete the survey.

- *Population.* In 1995, the NSF survey universe consisted of 722 responding units at 602 institutions. This is a complete survey universe and has been such since the fall of 1988. From 1984 to 1987, master's-degree-granting institutions were surveyed on a sample basis. During the fall 1988 survey cycle, the criteria for including departments in the survey universe were tightened, and all departments surveyed were reviewed. Departments not primarily oriented toward granting research degrees were no longer considered to meet the definition of S&E. As a result of the review, it was determined that a number of departments, primarily in the field of "Social Sciences, not elsewhere classified", were engaged in training primarily teachers, practitioners, administrators, or managers rather than researchers; these departments were deleted from the database. That process was continued during the fall 1989-1995 survey cycles and expanded to ensure trend consistency for the entire 1975-1995 period. As a result, total enrollments and social-science enrollments for all years were reduced. Any time-series problem between 1987 and 1988 should be small. The definition of "medical schools" was revised during the fall 1992 survey cycle to include only institutional components that are members of the Association of American Medical Colleges. That could effect data generated after the fall 1992 survey in that the association excludes schools of nursing, public health, dentistry, veterinary medicine, and other health-related disciplines; this change is not considered to have a major effect on the data.

- *Response Rate.* In 1995, 712 of 722 reporting units or 98.6%, were able to provide at least partial data. Of the 11,598 departments surveyed, 11,244 or 96.9%, responded. That is, 354 departments, or 3.1%, required complete imputation. Item nonresponse for the responding departments was 1,730, or 15.4 percent; these had one or more data cells imputed. Imputation for missing data elements was based on the prior year's data where available; otherwise, it was imputed on data on peer institutions.

ASSOCIATION OF AMERICAN MEDICAL COLLEGES MEDICAL FACULTY ROSTER SYSTEM

The Association of American Medical Colleges (AAMC) maintains several data bases that contain information on US medical personnel. One particularly relevant personnel system is AAMC's Medical Faculty Roster.

The Medical Faculty Roster is a comprehensive data directory of medical-school faculty, including education and employment history, nature of current activities, degrees, rank, and ethnicity. The data for this system are collected continuously from medical schools, as changes occur, through questionnaires that are completed by the faculty members. The accuracy of the data is considered to be very high, as was demonstrated by pilot samples for different studies conducted by AAMC. Data from this system can be linked to other data sources through Social Security numbers.

Appendix D

DOCTORAL FIELDS INCLUDED FOR DATA ANALYSIS

The Doctorate Records File (DRF; see appendix C) categorizes all fields in which PhDs are awarded. The committee has defined the life sciences as consisting primarily of the fields in DRF categories titled "agricultural sciences", "biological sciences", and "health sciences". Some fields in these categories have been excluded, for example, those in administrative, economic, or applied areas, such as agricultural economics. Two fields have been included as life sciences from engineering and chemistry categories and are listed below as "related sciences". Where the report refers to the "life sciences", it means all the fields listed below.

Where the committee distinguishes in the text, figures, and data tables between "biomedical" and "nonbiomedical" fields, it includes as nonbiomedical all the fields listed below in the agricultural sciences plus the 6 fields listed with an asterisk under "biological sciences". All other fields listed below are, in the committee's definition, biomedical fields.

Because the taxonomy of fields has changed over the last 30 years, explanations for changes in taxonomy are included.

Agricultural Sciences

Agronomy and Crop Science
Animal Breeding and Genetics: added in 1983
Animal Husbandry: dropped in 1983 and replaced with Animal Breeding and Genetics
Animal Nutrition
Animal Science, Other
Conservation/Renewable Natural Resources
Dairy Science
Fish and Wildlife: split into two categories in 1983
Fish Science and Management: added in 1983
Food Distribution: added in 1988; dropped again in 1995
Food Engineering: added in 1988
Food Science: split into three categories in 1988 but continues to appear on old forms
Food Science, Other: added in 1988
Forest Biology: added in 1988
Forest Engineering: added in 1988
Forest Management: added in 1988
Forestry and Related Science, Other: added in 1988
Forestry Science: split into several categories in 1988 but continues to appear on old forms
Horticulture Science
Plant Breeding and Genetics
Plant Pathology
Plant Protection and Pest Management: dropped in 1991 but continues to appear on old forms
Plant Sciences, Other
Poultry Science
Soil Chemistry/Microbiology: added in 1988
Soil Sciences: dropped in 1988 when split but continues to appear on old forms
Soil Sciences, Other: added in 1988
Wildlife: dropped in 1988 and replaced with Wildlife/Range Management but continues to appear on old forms.
Wildlife/Range Management: added in 1988
Wood Science and Pulp/Paper Technology: added in 1988
Agricultural Sciences, General
Agricultural Sciences, Other

Biological Sciences

Anatomy
Bacteriology: added in 1983

Biochemistry
Biometrics and Biostatistics
Biophysics
Biotechnology Research
* Botany
Cell Biology
Developmental Biology/Embryology
* Ecology
Endocrinology
* Entomology
Genetics, Animal and Plant: divided into two
    categories in 1983
Genetics, Human and Animal: added in 1983
Hydrobiology: dropped in 1980
Immunology
Microbiology: added in 1983
Microbiology and Bacteriology: split into two
    categories in 1983
Molecular Biology
Neuroscience
Nutritional Sciences
Parasitology
Pathology, Human and Animal
Pharmacology, Human and Animal

Physiology, Human and Animal
* Plant Genetics: added in 1983
* Plant Pathology
* Plant Physiology
Toxicology
Zoology
Biological Sciences, General
Biological Sciences, Other

Health Sciences

Environmental Health
Epidemiology: added in 1983
Pharmacy
Public Health: added in 1983
Public Health/Epidemiology: split into two
    categories in 1983
Health Sciences, General
Health Sciences, Other

Related Sciences

Bioengineering and Biomedical
Pharmaceutical Chemistry

Appendix E

DATA TABLES FOR CHAPTER 2

Appendix E

Table E.1 Demographic Characteristics of US Life-Science PhDs, 1963-1996

	1963	1964	1965	1966	1967	1968	1969	1970	1971	1972	1973	1974	1975	1976	1977	1978	1979
Total PhDs	2095	2356	2681	2887	3151	3695	4083	4503	4980	4855	4912	4734	4847	4800	4692	4809	4948
%	100	100	100	100	100	100	100	100	100	100	100	100	100	100	100	100	100
Men																	
Total	1887	2113	2401	2542	2729	3184	3517	3913	4265	4117	4046	3867	3888	3835	3769	3754	3810
%	90.1	89.7	89.6	88.0	86.6	86.2	86.1	86.9	85.6	84.8	82.4	81.7	80.2	79.9	80.3	78.1	77.0
US citizen/perm. residents	1582	1706	1888	2024	2240	2607	2946	3337	3656	3511	3457	3100	3235	3220	3151	3163	3207
%	75.5	72.4	70.4	70.1	71.1	70.6	72.2	74.1	73.4	72.3	70.4	65.5	66.7	67.1	67.2	65.8	64.8
Women																	
Total	208	243	280	345	422	511	566	590	715	738	866	867	959	965	923	1055	1138
%	9.9	10.3	10.4	12.0	13.4	13.8	13.9	13.1	14.4	15.2	17.6	18.3	19.8	20.1	19.7	21.9	23.0
US citizen/perm. residents	172	199	245	291	349	460	498	525	637	670	783	752	886	855	804	919	1014
%	8.2	8.4	9.1	10.1	11.1	12.4	12.2	11.7	12.8	13.8	15.9	15.9	18.3	17.8	17.1	19.1	20.5
Whites (all PhDs)											3104	3521	3838	3790	3718	3661	3798
%											63.2	74.4	79.2	79.0	79.2	76.1	76.8
Whites (US and perms)											2910	3284	3628	3591	3495	3473	3639
%											59.2	69.4	74.9	74.8	74.5	72.2	73.5
Minorities (US & perms)											96	96	104	105	109	134	110
%											2.0	2.0	2.1	2.2	2.3	2.8	2.2
Total US & perms	1754	1905	2133	2315	2589	3067	3444	3862	4293	4181	4240	3852	4121	4075	3955	4082	4221
%	83.7	80.9	79.6	80.2	82.2	83.0	84.3	85.8	86.2	86.1	86.3	81.4	85.0	84.9	84.3	84.9	85.3

Appendix E

Table E.1 (cont'd)

	1963	1964	1965	1966	1967	1968	1969	1970	1971	1972	1973	1974	1975	1976	1977	1978	1979
Temp visas staying in US	103	104	125	157	174	186	191	178	147	128	137	169	175	160	166	156	147
%	4.9	4.4	4.7	5.4	5.5	5.0	4.7	4.0	3.0	2.6	2.8	3.6	3.6	3.3	3.5	3.2	3.0
Temp visas leaving US	168	229	280	262	260	301	284	324	336	345	342	367	334	318	327	325	335
%	8.0	9.7	10.4	9.1	8.3	8.1	7.0	7.2	6.7	7.1	7.0	7.8	6.9	6.6	7.0	6.8	6.8
Total temporary residents	325	415	518	525	517	586	589	606	589	574	584	663	628	603	623	602	606
%	15.5	17.6	19.3	18.2	16.4	15.9	14.4	13.5	11.8	11.8	11.9	14.0	13.0	12.6	13.3	12.5	12.2
Postdoctoral appointments	485	567	709	764	873	1095	1305	1607	1729	1720	1797	1655	1970	2046	2026	2161	2274
%	23.2	24.1	26.4	26.5	27.7	29.6	32.0	35.7	34.7	35.4	36.6	35.0	40.6	42.6	43.2	44.9	46.0
Elapsed time to degree								6.0	6.0	6.2	6.3	6.3	6.2	6.4	6.5	6.5	6.6
Median age at time of degree								29.3	29.4	29.8	30.1	30.1	29.8	29.8	29.9	30.0	29.9

Appendix E

Table E.1 Demographic Characteristics of US Life-Science PhDs, 1963-1996

	1980	1981	1982	1983	1984	1985	1986	1987	1988	1989	1990	1991	1992	1993	1994	1995	1996	Total
Total PhDs	5180	5288	5362	5263	5414	5428	5360	5399	5807	5908	6211	6508	6682	6924	7182	7312	7696	171952
%	100	100	100	100	100	100	100	100	100	100	100	100	100	100	100	100	100	100
Men																		
Total	3909	3936	3923	3737	3855	3806	3688	3649	3844	3842	4058	4183	4231	4247	4377	4425	4552	125899
%	75.5	74.4	73.2	71.0	71.2	70.1	68.8	67.6	66.2	65.0	65.3	64.3	63.3	61.3	60.9	60.5	59.1	73.2
US citizen/perm. residents	3294	3287	3243	3040	3118	2993	2841	2752	2814	2785	2866	2924	2813	2852	3142	3241	3139	99174
%	63.6	62.2	60.5	57.8	57.6	55.1	53.0	51.0	48.5	47.1	46.1	44.9	42.1	41.2	43.7	44.3	40.8	57.7
Women																		
Total	1271	1352	1439	1526	1559	1622	1672	1750	1963	2066	2153	2301	2420	2632	2777	2848	3113	45855
%	24.5	25.6	26.8	29.0	28.8	29.9	31.2	32.4	33.8	35.0	34.7	35.4	36.2	38.0	38.7	38.9	40.4	26.7
US citizen/perm. Residents	1137	1200	1289	1363	1374	1380	1428	1447	1603	1674	1701	1790	1872	2001	2236	2319	2440	38313
%	21.9	22.7	24.0	25.9	25.4	25.4	26.6	26.8	27.6	28.3	27.4	27.5	28.0	28.9	31.1	31.7	31.7	22.3
Whites (All PhDs)	4031	4118	4259	4155	4211	4126	3995	3897	4155	4177	4331	4391	4348	4436	4476	4410	4518	97464
%	77.8	77.9	79.4	78.9	77.8	76.0	74.5	72.2	71.6	70.7	69.7	67.5	65.1	64.1	62.3	60.3	58.7	56.7
Whites (US & perms)	3829	3929	4043	3931	3995	3888	3778	3667	3889	3911	3994	4043	4004	4070	4099	4048	4078	91216
%	73.9	74.3	75.4	74.7	73.8	71.6	70.5	67.9	67.0	66.2	64.3	62.1	59.9	58.8	57.1	55.4	53.0	53.0
Minorities (US & perms)	115	143	142	130	158	178	186	192	210	206	221	250	242	274	319	340	341	4401
%	2.2	2.7	2.6	2.5	2.9	3.3	3.5	3.6	3.6	3.5	3.6	3.8	3.6	4.0	4.4	4.6	4.4	2.6
Total US & Perms	4431	4487	4532	4403	4492	4373	4269	4199	4417	4459	4567	4714	4685	4853	5378	5561	5579	137488
%	85.5	84.9	84.5	83.7	83.0	80.6	79.6	77.8	76.1	75.5	73.5	72.4	70.1	70.1	74.9	76.1	72.5	80.0

Appendix E

Table E.1 (cont'd)

	1980	1981	1982	1983	1984	1985	1986	1987	1988	1989	1990	1991	1992	1993	1994	1995	1996	Total
Temp visas staying in US	167	156	176	191	223	263	264	333	413	476	606	928	1132	1165	960	910	1196	12162
%	3.2	3.0	3.3	3.6	4.1	4.8	4.9	6.2	7.1	8.1	9.8	14.3	16.9	16.8	13.4	12.4	15.5	7.1
Temp visas leaving US	345	358	353	356	344	408	353	352	376	372	571	597	633	658	658	624	650	13145
%	6.7	6.8	6.6	6.8	6.4	7.5	6.6	6.5	6.5	6.3	9.2	9.2	9.5	9.5	9.2	8.5	8.4	7.6
Total temporary residents	650	663	674	719	749	865	806	887	1032	1102	1472	1683	1886	1927	1726	1633	1939	29966
%	12.5	12.5	12.6	13.7	13.8	15.9	15.0	16.4	17.8	18.7	23.7	25.9	28.2	27.8	24.0	22.3	25.5	17.4
Postdoctoral appointments	2425	2395	2474	2400	2491	2486	2588	2647	2946	2949	3162	3371	3497	3656	3731	3768	3940	75709
%	46.8	45.3	46.1	45.6	46.0	45.8	48.3	49.0	50.7	49.9	50.9	51.8	52.3	52.8	51.9	51.5	51.2	44.0
Elapsed time to degree	6.6	6.7	6.9	7.0	7.3	7.3	7.5	7.6	7.7	7.7	7.9	7.9	8.0	8.0	8.0	8.0	8.0	
Median age at time of degree	29.9	29.8	30.2	30.4	30.8	31.1	31.3	31.4	31.7	31.8	32.0	32.1	32.2	32.0	32.0	32.0	32.0	

Appendix E

Table E.2 Demographic Characteristics of US Life-Science PhDs in the Biomedical Sciences, 1963-1996

	1963	1964	1965	1966	1967	1968	1969	1970	1971	1972	1973	1974	1975	1976	1977	1978	1979
Total PhDs	1319	1510	1717	1916	2141	2501	2817	3136	3449	3402	3437	3336	3408	3483	3393	3449	3560
%	100	100	100	100	100	100	100	100	100	100	100	100	100	100	100	100	100
Men																	
Total	1136	1311	1472	1614	1778	2045	2325	2622	2828	2753	2704	2578	2563	2648	2606	2542	2589
%	86.1	86.8	85.7	84.2	83.0	81.8	82.5	83.6	82.0	80.9	78.7	77.3	75.2	76.0	76.8	73.7	72.7
US citizen /perm. residents	975	1098	1206	1342	1525	1753	2027	2308	2505	2444	2415	2180	2250	2324	2298	2253	2308
%	73.9	72.7	70.2	70.0	71.2	70.1	72.0	73.6	72.6	71.8	70.3	65.3	66.0	66.7	67.7	65.3	64.8
Women																	
Total	183	199	245	302	363	456	492	514	621	649	733	758	845	835	787	907	971
%	13.9	13.2	14.3	15.8	17.0	18.2	17.5	16.4	18.0	19.1	21.3	22.7	24.8	24.0	23.2	26.3	27.3
US citizen /perm. residents	154	170	221	261	312	413	441	463	566	599	672	676	782	747	697	810	880
%	11.7	11.3	12.9	13.6	14.6	16.5	15.7	14.8	16.4	17.6	19.6	20.3	22.9	21.4	20.5	23.5	24.7
Whites (all PhDs)											2189	2513	2744	2786	2752	2668	2801
%											63.7	75.3	80.5	80.0	81.1	77.4	78.7
Whites (US & perms)											2088	2404	2641	2677	2629	2579	2727
%											60.8	72.1	77.5	76.9	77.5	74.8	76.6
Minorities (US & perms)											78	81	83	81	76	101	84
%											2.3	2.4	2.4	2.3	2.2	2.9	2.4

Appendix E

Table E.2 (Cont'd)

	1963	1964	1965	1966	1967	1968	1969	1970	1971	1972	1973	1974	1975	1976	1977	1978	1979
Total US & perms	1129	1268	1427	1603	1837	2166	2468	2771	3071	3043	3087	2856	3032	3071	2995	3063	3188
%	85.6	84.0	83.1	83.7	85.8	86.6	87.6	88.4	89.0	89.4	89.8	85.6	89.0	88.2	88.3	88.8	89.6
Temp visas staying in US	77	80	91	119	125	131	129	128	101	94	94	121	120	130	114	121	115
%	5.8	5.3	5.3	6.2	5.8	5.2	4.6	4.1	2.9	2.8	2.7	3.6	3.5	3.7	3.4	3.5	3.2
Temp visas leaving US	76	97	123	103	95	110	129	146	123	128	124	123	114	113	121	108	110
%	5.8	6.4	7.2	5.4	4.4	4.4	4.6	4.7	3.6	3.8	3.6	3.7	3.3	3.2	3.6	3.1	3.1
Total temporary residents	181	214	273	279	269	304	309	333	284	269	269	307	285	301	297	284	272
%	13.7	14.2	15.9	14.6	12.6	12.2	11.0	10.6	8.2	7.9	7.8	9.2	8.4	8.6	8.8	8.2	7.6
Postdoctoral appointments	404	483	589	668	759	949	1121	1406	1495	1516	1552	1459	1756	1833	1826	1938	2030
%	30.6	32.0	34.3	34.9	35.5	37.9	39.8	44.8	43.3	44.6	45.2	43.7	51.5	52.6	53.8	56.2	57.0
Time to degree								6.0	6.0	6.0	6.2	6.3	6.1	6.3	6.3	6.4	6.5
Median age at time of degree								29.1	29.1	29.4	29.7	29.6	29.4	29.4	29.4	29.6	29.6

Appendix E

Table E.2 Demographic Characteristics of US Life-Science PhDs in the Biomedical Sciences, 1963-1996

	1980	1981	1982	1983	1984	1985	1986	1987	1988	1989	1990	1991	1992	1993	1994	1995	1996	Total
Total PhDs	3742	3750	3866	3693	3813	3698	3789	3915	4262	4318	4501	4865	5060	5430	5523	5728	6021	123948
%	100	100	100	100	100	100	100	100	100	100	100	100	100	100	100	100	100	100
Men																		
Total	2677	2650	2698	2433	2531	2395	2426	2467	2637	2616	2743	2925	3030	3170	3150	3245	3383	85290
%	71.5	70.7	69.8	65.9	66.4	64.8	64.0	63.0	61.9	60.6	60.9	60.1	59.9	58.4	57.0	56.7	56.2	68.8
US citizen/perm. residents	2403	2371	2336	2110	2167	1998	1978	1941	2036	2005	2043	2152	2133	2254	2403	2493	2445	70479
%	64.2	63.2	60.4	57.1	56.8	54.0	52.2	49.6	47.8	46.4	45.4	44.2	42.2	41.5	43.5	43.5	40.6	56.9
Women																		
Total	1065	1100	1168	1260	1282	1303	1363	1448	1625	1702	1758	1921	2008	2220	2347	2452	2609	38491
%	28.5	29.3	30.2	34.1	33.6	35.2	36.0	37.0	38.1	39.4	39.1	39.5	39.7	40.9	42.5	42.8	43.3	31.1
US citizen/perm. residents	962	1003	1058	1138	1150	1132	1185	1216	1353	1399	1416	1521	1591	1732	1936	2049	2082	32787
%	25.7	26.7	27.4	30.8	30.2	30.6	31.3	31.1	31.7	32.4	31.5	31.3	31.4	31.9	35.1	35.8	34.6	26.5
Whites (All PhDs)	2978	3025	3113	2974	3020	2872	2886	2879	3122	3116	3216	3337	3363	3525	3496	3472	3519	72366
%	79.6	80.7	80.5	80.5	79.2	77.7	76.2	73.5	73.3	72.2	71.5	68.6	66.5	64.9	63.3	60.6	58.4	58.4
Whites (US & perms)	2879	2946	2996	2872	2914	2754	2777	2736	2979	2963	3006	3113	3158	3304	3259	3229	3231	68861
%	76.9	78.6	77.5	77.8	76.4	74.5	73.3	69.9	69.9	68.6	66.8	64.0	62.4	60.8	59.0	56.4	53.7	55.6
Minorities (US & perms)	87	104	116	104	121	126	135	150	151	161	165	199	191	220	256	277	273	3420
%	2.3	2.8	3.0	2.8	3.2	3.4	3.6	3.8	3.5	3.7	3.7	4.1	3.8	4.1	4.6	4.8	4.5	
Total US & perms	3365	3374	3394	3248	3317	3130	3163	3157	3389	3404	3459	3673	3724	3986	4339	4543	4527	103267
%	89.9	90.0	87.8	88.0	87.0	84.6	83.5	80.6	79.5	78.8	76.8	75.5	73.6	73.4	78.6	79.3	75.2	83.3

Appendix E

Table E.2 (cont'd)

	1980	1981	1982	1983	1984	1985	1986	1987	1988	1989	1990	1991	1992	1993	1994	1995	1996	Total
Temp visas staying in US	130	112	130	155	164	205	193	278	327	368	473	732	893	944	743	704	949	9390
%	3.5	3.0	3.4	4.2	4.3	5.5	5.1	7.1	7.7	8.5	10.5	15.0	17.6	17.4	13.5	12.3	15.8	7.6
Temp visas leaving US	108	97	146	100	108	122	126	128	150	155	267	281	287	310	303	313	334	5278
%	2.9	2.6	3.8	2.7	2.8	3.3	3.3	3.3	3.5	3.6	5.9	5.8	5.7	5.7	5.5	5.5	5.5	4.3
Total temporary residents	299	278	341	335	355	417	417	523	596	662	922	1108	1252	1317	1118	1084	1341	17095
%	8.0	7.4	8.8	9.1	9.3	11.3	11.0	13.4	14.0	15.3	20.5	22.8	24.7	24.3	20.2	18.9	22.3	13.8
Postdoctoral appointments	2159	2118	2169	2098	2125	2095	2190	2255	2498	2502	2665	2872	3009	3204	3228	3319	3454	65744
%	57.7	56.5	56.1	56.8	55.7	56.7	57.8	57.6	58.6	57.9	59.2	59.0	59.5	59.0	58.4	57.9	57.4	53.0
Time to degree	6.5	6.6	6.9	7.0	7.3	7.3	7.3	7.5	7.6	7.5	7.7	7.7	7.8	7.9	7.9	7.9	7.8	
Median age at time of degree	29.4	29.5	29.8	30.0	30.5	30.7	30.9	30.8	31.2	31.3	31.4	31.4	31.6	31.4	31.4	31.5	31.5	

114

Appendix E

Table E.3 Demographic Characteristics of US Life-Science PhDs in Nonbiomedical Life Sciences, 1963-1996

	1963	1964	1965	1966	1967	1968	1969	1970	1971	1972	1973	1974	1975	1976	1977	1978	1979
Total PhDs	776	846	964	971	1010	1194	1266	1367	1531	1453	1475	1398	1439	1317	1299	1360	1388
%	100	100	100	100	100	100	100	100	100	100	100	100	100	100	100	100	100
Men																	
Total	751	802	929	928	951	1139	1192	1291	1437	1364	1342	1289	1325	1187	1163	1212	1221
%	96.8	94.8	96.4	95.6	94.2	95.4	94.2	94.4	93.9	93.9	91.0	92.2	92.1	90.1	89.5	89.1	88.0
US citizen/perm. residents	607	608	682	682	715	854	919	1029	1151	1067	1042	920	985	896	853	910	899
%	78.2	71.9	70.7	70.2	70.8	71.5	72.6	75.3	75.2	73.4	70.6	65.8	68.5	68.0	65.7	66.9	64.8
Women																	
Total	25	44	35	43	59	55	74	76	94	89	133	109	114	130	136	148	167
%	3.2	5.2	3.6	4.4	5.8	4.6	5.8	5.6	6.1	6.1	9.0	7.8	7.9	9.9	10.5	10.9	12.0
US citizen/prm. residents	18	29	24	30	37	47	57	62	71	71	111	76	104	108	107	109	134
%	2.3	3.4	2.5	3.1	3.7	3.9	4.5	4.5	4.6	4.9	7.5	5.4	7.2	8.2	8.2	8.0	9.7
Whites (all PhDs)											915	1008	1094	1004	966	993	997
%											62.0	72.1	76.0	76.2	74.4	73.0	71.8
Whites (US & perms)											822	880	987	914	866	894	912
%											55.7	62.9	68.6	69.4	66.7	65.7	65.7
Minorities (US & perms)											18	15	21	24	33	33	26
%											1.2	1.1	1.5	1.8	2.5	2.4	1.9
Total US & perms	625	637	706	712	752	901	976	1091	1222	1138	1153	996	1089	1004	960	1019	1033
%	80.5	75.3	73.2	73.3	74.5	75.5	77.1	79.8	79.8	78.3	78.2	71.2	75.7	76.2	73.9	74.9	74.4

Appendix E

Table E.3 (cont'd)

	1963	1964	1965	1966	1967	1968	1969	1970	1971	1972	1973	1974	1975	1976	1977	1978	1979
Temp visas staying in US	26	24	34	38	49	55	62	50	46	34	43	48	55	30	52	35	32
%	3.4	2.8	3.5	3.9	4.9	4.6	4.9	3.7	3.0	2.3	2.9	3.4	3.8	2.3	4.0	2.6	2.3
Temp visas leaving US	92	132	157	159	165	191	155	178	213	217	218	244	220	205	206	217	225
%	11.9	15.6	16.3	16.4	16.3	16.0	12.2	13.0	13.9	14.9	14.8	17.5	15.3	15.6	15.9	16.0	16.2
Total temporary residents	144	201	245	246	248	282	280	273	305	305	315	356	343	302	326	318	334
%	18.6	23.8	25.4	25.3	24.6	23.6	22.1	20.0	19.9	21.0	21.4	25.5	23.8	22.9	25.1	23.4	24.1
Postdoctoral appointments	81	84	120	96	114	146	184	201	234	204	245	196	214	213	200	223	244
%	10.4	9.9	12.4	9.9	11.3	12.2	14.5	14.7	15.3	14.0	16.6	14.0	14.9	16.2	15.4	16.4	17.6
Time to degree								6.0	6.1	6.3	6.5	6.4	6.5	6.6	6.9	6.9	6.8
Median age at time of degree								29.9	30.4	30.7	31.1	31.0	31.0	31.0	31.0	31.1	31.0

Appendix E

Table E.3 Demographic Characteristics of US Life-Science PhDs in Nonbiomedical Life Sciences, 1963-1996

	1980	1981	1982	1983	1984	1985	1986	1987	1988	1989	1990	1991	1992	1993	1994	1995	1996	Total
Total PhDs	1438	1538	1496	1570	1601	1730	1571	1484	1545	1590	1710	1643	1622	1494	1659	1584	1675	48004
%	100	100	100	100	100	100	100	100	100	100	100	100	100	100	100	100	100	100
Men																		
Total	1232	1286	1225	1304	1324	1411	1262	1182	1207	1226	1315	1258	1201	1077	1227	1180	1169	40609
%	85.7	83.6	81.9	83.1	82.7	81.6	80.3	79.6	78.1	77.1	76.9	76.6	74.0	72.1	74.0	74.5	69.8	84.6
US citizen/perm. residents	891	916	907	930	951	995	863	811	778	780	823	772	680	598	739	748	694	28695
%	62.0	59.6	60.6	59.2	59.4	57.5	54.9	54.6	50.4	49.1	48.1	47.0	41.9	40.0	44.5	47.2	41.4	59.8
Women																		
Total	206	252	271	266	277	319	309	302	338	364	395	380	412	412	430	396	504	7364
%	14.3	16.4	18.1	16.9	17.3	18.4	19.7	20.4	21.9	22.9	23.1	23.1	25.4	27.6	25.9	25.0	30.1	15.3
US citizen/perm. residents	175	197	231	225	224	248	243	231	250	275	285	269	281	269	300	270	358	5526
%	12.2	12.8	15.4	14.3	14.0	14.3	15.5	15.6	16.2	17.3	16.7	16.4	17.3	18.0	18.1	17.0	21.4	11.5
Whites (all PhDs)	1053	1093	1146	1181	1191	1254	1109	1018	1033	1061	1115	1054	985	911	980	938	999	25098
%	73.2	71.1	76.6	75.2	74.4	72.5	70.6	68.6	66.9	66.7	65.2	64.2	60.7	61.0	59.1	59.2	59.6	52.3
Whites (US & perms)	950	983	1047	1059	1081	1134	1001	931	910	948	988	930	846	766	840	819	847	22355
%	66.1	63.9	70.0	67.5	67.5	65.5	63.7	62.7	58.9	59.6	57.8	56.6	52.2	51.3	50.6	51.7	50.6	46.6
Minorities (US & perms)	28	39	26	26	37	52	51	42	59	45	56	51	51	54	63	63	68	981
%	1.9	2.5	1.7	1.7	2.3	3.0	3.2	2.8	3.8	2.8	3.3	3.1	3.1	3.6	3.8	4.0	4.1	2.0
Total US & perms	1066	1113	1138	1155	1175	1243	1106	1042	1028	1055	1108	1041	961	867	1039	1018	1052	34221
%	74.1	72.4	76.1	73.6	73.4	71.8	70.4	70.2	66.5	66.4	64.8	63.4	59.2	58.0	62.6	64.3	62.8	71.3

Appendix E

Table E.3 (cont'd)

	1980	1981	1982	1983	1984	1985	1986	1987	1988	1989	1990	1991	1992	1993	1994	1995	1996	Total
Temp visas staying in US	37	44	46	36	59	58	71	55	86	108	133	196	239	221	217	206	247	2772
%	2.6	2.9	3.1	2.3	3.7	3.4	4.5	3.7	5.6	6.8	7.8	11.9	14.7	14.8	13.1	13.0	14.7	5.8
Temp visas leaving US	237	261	207	256	236	286	227	224	226	217	304	316	346	348	355	311	316	7867
%	16.5	17.0	13.8	16.3	14.7	16.5	14.4	15.1	14.6	13.6	17.8	19.2	21.3	23.3	21.4	19.6	18.9	16.4
Total temporary residents	351	385	333	384	394	448	389	364	436	440	550	575	634	610	608	549	598	12871
%	24.4	25.0	22.3	24.5	24.6	25.9	24.8	24.5	28.2	27.7	32.2	35.0	39.1	40.8	36.6	34.7	35.7	26.8
Postdoctoral appointments	266	277	305	302	366	391	398	392	447	447	497	499	488	452	503	449	486	9964
%	18.5	18.0	20.4	19.2	22.9	22.6	25.3	26.4	28.9	28.1	29.1	30.4	30.1	30.3	30.3	28.3	29.0	20.8
Time to degree	6.9	6.9	7.0	7.0	7.3	7.3	7.6	7.7	8.0	8.1	8.2	8.3	8.3	8.4	8.5	8.4	8.6	
Median age at time of degree	31.1	31.0	31.0	31.3	31.6	31.9	32.3	32.7	32.9	33.3	33.6	33.8	34.0	34.0	34.3	34.2	34.1	

Appendix E

Table E.4 Demographic and Other US Life-Science PhDs by Fine Field in 10- or 11-Year Cohorts

	FY 86-96 Doctorate Recipients				FY 76-85 Doctorate Recipients					FY 66-75 Doctorate Recipients				
	Total PhDs n	Women %	Temp. Res. %	Planning Postdoc %	Median Time to Degree yrs	Total PhDs n	Women %	Temp. Res. %	Planning Postdoc %	Median Time to Degree yrs	Total PhDs n	Women %	Temp. Res. %	Planning Postdoc %	Median Time to Degree yrs**
Nonbiomedical doctoral fields															
Agronomy and crop science	1437	13	43	20	7.9	1475	7	39	12	6.6	1636	1	33	9	6.2
Animal breeding and genetics	227	15	41	24	7.1	81	21	31	14	6.5					
Animal husbandry	634	19	27	22	7.0	155	5	19	16	6.3	865	1	22	12	6.0
Animal nutrition	962	22	31	31	7.2	1039	10	28	13	6.3	448	6	27	13	6.2
Animal science, other	1201	38	20	40	8.6	277	13	31	21	6.7					
Botany	134	19	33	13	9.9	1428	27	9	28	7.2	1784	17	10	17	6.5
Conservation/renewable natural research															
Dairy science	126	14	33	27	7.5										
Fish and wildlife	427	16	27	18	9.2	452	7	10	9	7.3	437	2	7	7	6.5
Fish science and management	1	0	0	0	6.8	117	14	17	10	7.6					
Food distribution	92	13	59	22	8.0										
Food engineering	269	34	43	19	8.0	1128	25	37	12	6.9	538	13	29	14	6.6
Food science	1265	39	51	20	8.0										
Food science, other	197	24	27	30	8.5										
Forest biology	17	6	47	12	10.5										
Forest engineering	167	12	35	13	8.6										
Forest management	502	17	34	14	9.2										
Forestry and related science, other	203	12	23	20	7.7	862	6	21	9	7.2	730	0	19	6	6.6
Forestry science	778	25	38	19	8.0	704	15	31	8	6.6	665	3	32	7	6.2
Horticulture science	817	18	41	25	7.6	237	15	27	17	6.2					
Plant breeding and genetics															

118

Appendix E

Table E.4 (cont'd)

	FY 86-96 Doctorate Recipients					FY 76-85 Doctorate Recipients					FY 66-75 Doctorate Recipients				
	Total PhDs n	Women %	Temp. Res. %	Planning Postdoc %	Median Time to Degree yrs	Total PhDs n	Women %	Temp. Res. %	Planning Postdoc %	Median Time to Degree yrs	Total PhDs n	Women %	Temp. Res. %	Planning Postdoc %	Median Time to Degree yrs**
Plant pathology	742	29	44	36	8.0	911	17	29	23	6.5	987	4	31	19	6.3
Plant protection and pest management	13	8	31	23	7.1										
Plant science, other	252	21	39	29	8.0	57	11	28	12	6.7					
Poultry science	133	21	39	20	7.6										
Soil chemistry/microbiology	239	22	36	30	8.3										
Soil science	195	9	36	24	8.0	842	7	44	15	7.2	512	2	43	16	6.7
Soil science, other	647	14	44	25	8.8										
Wildlife	46	7	9	11	7.3	100	5	13	9	7.5					
Wildlife/range management	480	21	20	18	9.0										
Wood sci and pulp/paper technology	166	13	55	15	8.5										
Agricultural sciences, general	67	19	57	18	8.8	54	11	56	6	8.5	60	2	47	7	6.8
Agricultural sciences, other	268	18	40	16	8.2	754	8	33	10	7.0	852	3	29	8	6.2
Biomedical doctoral fields															
Anatomy	821	40	14	62	8.0	1355	31	5	62	6.8	1163	21	5	36	6.0
Bacteriology	143	40	16	62	8.0	39	31	8	67	6.8					
Biochemistry	7856	38	22	72	7.0	6237	27	9	72	6.3	5772	20	12	62	5.8
Biomedical siences	233	41	18	59	8.8										
Biometrics & biostatistics	623	39	27	13	8.6	470	29	14	8	7.3	308	19	16	7	6.7
Biophysics	1193	25	26	71	7.4	1052	15	11	65	6.7	1062	10	10	56	6.0
Biotechnology research	32	28	28	28	7.6										
Cell biology	1927	45	14	75	7.6	628	35	5	72	7.0	414	37	7	49	6.0
Developmental Biology/embryology	418	49	16	82	7.6	152	39	5	76	6.5	340	32	4	51	5.9

Appendix E

Table E.4 (cont'd)

	FY 86-96 Doctorate Recipients					FY 76-85 Doctorate Recipients					FY 66-75 Doctorate Recipients				
	Total PhDs n	Women %	Temp. Res. %	Planning Postdoc %	Median Time to Degree yrs	Total PhDs n	Women %	Temp. Res. %	Planning Postdoc %	Median Time to Degree yrs	Total PhDs n	Women %	Temp. Res. %	Planning Postdoc %	Median Time to Degree yrs**
Ecology	2018	31	14	37	9.0	1771	20	6	28	7.3	1047	9	8	16	6.1
Endocrinology	248	46	23	65	8.0	75	39	11	73	7.0					
Entomology	1483	21	25	32	8.5	1550	13	20	22	7.1	1726	6	18	16	6.3
Genetics, animal & plant						1041	35	12	59	6.5	1409	21	22	39	6.0
Genetics, human & animal	1678	46	16	71	7.5	282	46	6	67	7.3					
Hydrobiology						40	15	0	25	7.0	161	5	4	17	6.7
Immunology	1882	45	13	74	7.5	1257	37	7	73	6.5	208	27	7	62	6.0
Microbiology	4110	39	20	67	7.7	944	37	10	64	7.0					
Microbiology & bacteriology						2416	29	8	62	6.3	3695	22	9	46	6.1
Molecular biology	5247	40	20	78	7.5	1961	33	8	79	6.5	864	29	8	74	5.7
Neurosciences	2558	39	13	78	7.5	552	34	6	79	6.9					
Nutritional sciences	1433	70	23	33	8.5	1006	61	16	30	7.1					
Parasitology	206	36	22	50	8.4	184	16	16	39	7.5	63	19	10	37	7.1
Pathology, human & animal	1274	36	18	49	8.3	974	26	11	42	7.0	605	8	13	25	6.9
Pharmacology, human & animal	2889	39	16	71	7.0	2344	24	7	67	6.2	1537	14	9	50	6.0
Physiology, human & animal	2898	36	18	63	7.8	2939	23	7	66	6.6	3087	16	8	46	6.0
Plant genetics	324	28	31	65	7.6	70	24	24	50	7.7					
Plant pathology	383	30	37	50	8.0	97	22	22	36	6.3					
Plant Physiology	665	32	28	53	8.1	576	23	17	49	6.8	817	13	22	31	6.2
Toxicology	1204	36	12	45	7.9	256	25	5	48	7.2					
Zoology	1450	30	13	40	8.7	2112	23	4	34	7.5	3008	16	5	26	6.3
Biological sciences, general	3087	39	20	50	8.0	1908	31	8	43	7.0	1183	24	8	31	6.0
Biological sciences, other	1560	37	17	49	8.3	1416	31	7	57	6.9	1405	29	11	35	6.2
Environmental health	473	37	22	20	8.9	356	24	10	19	7.6	71	17	0	13	6.5
Epidemiology	1285	59	17	20	9.3	255	59	14	15	8.3					
Pharmacy	1417	34	39	22	7.8	762	19	19	17	7.0	559	8	18	13	6.2
Public health	1464	63	18	11	9.8	219	59	9	5	8.8					
Public health/epidemiology						887	44	10	9	7.8	792	25	10	4	7.3

Appendix E

Table E.4 (cont'd)

	FY 86-96 Doctorate Recipients					FY 76-85 Doctorate Recipients					FY 66-75 Doctorate Recipients				
	Total PhDs n	Women %	Temp. Res. %	Planning Postdoc %	Median Time to Degree yrs	Total PhDs n	Women %	Temp. Res. %	Planning Postdoc %	Median Time to Degree yrs	Total PhDs n	Women %	Temp. Res. %	Planning Postdoc %	Median Time to Degree yrs**
Health science, general	317	43	15	30	8.1	167	30	7	44	7.0	125	18	7	31	6.0
Health sciences, other	1084	46	17	16	9.1	909	31	10	28	7.4	701	31	9	23	6.9
Bioengineering and biomedical	1549	21	27	36	7.3	700	6	13	31	7.0	439	2	8	29	6.3
Pharmaceutical chemistry	853	32	28	42	7.4	552	18	16	41	6.5	572	8	17	30	6.5

* Total for 1973-75. Data on race/ethnicity is available only for 1973 to date. Percentages calculated with base years 1973 to 1975.

** Median computed only for period 1970 to 1975

Appendix E

Table E.5 Demographic Characteristics of US Life-Science PhDs by Citizenship, 1963-1996

	1963	1964	1965	1966	1967	1968	1969	1970	1971	1972	1973	1974	1975	1976	1977	1978	1979	1980
Total PhDs	2095	2356	2681	2887	3151	3695	4083	4503	4980	4855	4912	4734	4847	4800	4692	4809	4948	5180
%	100	100	100	100	100	100	100	100	100	100	100	100	100	100	100	100	100	100
US citizen & prm. residents	1754	1905	2133	2315	2589	3067	3444	3862	4293	4181	4240	3852	4121	4075	3955	4082	4221	4431
%	83.7	80.9	79.6	80.2	82.2	83.0	84.3	85.8	86.2	86.1	86.3	81.4	85.0	84.9	84.3	84.9	85.3	85.5
US citizens–native born	1692	1824	2039	2213	2392	2783	3152	3504	3854	3705	3746	3405	3673	3684	3593	3676	3872	4054
%	80.8	77.4	76.1	76.7	75.9	75.3	77.2	77.8	77.4	76.3	76.3	71.9	75.8	76.8	76.6	76.4	78.3	78.3
US citizen–naturalized				4	64	100	92	109	118	113	128	117	136	142	122	175	141	150
%	0.0	0.0	0.0	0.1	2.0	2.7	2.3	2.4	2.4	2.3	2.6	2.5	2.8	3.0	2.6	3.6	2.8	2.9
US citizens	1692	1824	2039	2217	2456	2883	3244	3613	3972	3818	3874	3522	3809	3826	3715	3851	4013	4204
%	80.8	77.4	76.1	76.8	77.9	78.0	79.5	80.2	79.8	78.6	78.9	74.4	78.6	79.7	79.2	80.1	81.1	81.2
Permanent residents	62	81	94	98	133	184	200	249	321	363	366	330	312	249	240	231	208	227
%	3.0	3.4	3.5	3.4	4.2	5.0	4.9	5.5	6.4	7.5	7.5	7.0	6.4	5.2	5.1	4.8	4.2	4.4
Temp visas staying in US	103	104	125	157	174	186	191	178	147	128	137	169	175	160	166	156	147	167
%	4.9	4.4	4.7	5.4	5.5	5.0	4.7	4.0	3.0	2.6	2.8	3.6	3.6	3.3	3.5	3.2	3.0	3.2
Temp visas leaving US	168	229	280	262	260	301	284	324	336	345	342	367	334	318	327	325	335	345
%	8.0	9.7	10.4	9.1	8.3	8.1	7.0	7.2	6.7	7.1	7.0	7.8	6.9	6.6	7.0	6.8	6.8	6.7
Total temporary residents	325	415	518	525	517	586	589	606	589	574	584	663	628	603	623	602	606	650
%	15.5	17.6	19.3	18.2	16.4	15.9	14.4	13.5	11.8	11.8	11.9	14.0	13.0	12.6	13.3	12.5	12.2	12.5

Appendix E

Table E.5 (cont'd)

	1963	1964	1965	1966	1967	1968	1969	1970	1971	1972	1973	1974	1975	1976	1977	1978	1979	1980
Non responses	16	36	30	47	45	42	50	35	98	100	88	219	98	122	114	125	121	99
%	0.8	1.5	1.1	1.6	1.4	1.1	1.2	0.8	2.0	2.1	1.8	4.6	2.0	2.5	2.4	2.6	2.4	1.9

# Appendix E

Table E.5 Demographic Characteristics of US Life-Science PhDs by Citizenship, 1963-1996

	1981	1982	1983	1984	1985	1986	1987	1988	1989	1990	1991	1992	1993	1994	1995	1996
Total PhDs	5288	5362	5263	5414	5428	5360	5399	5807	5908	6211	6508	6682	6924	7182	7312	7696
%	100	100	100	100	100	100	100	100	100	100	100	100	100	100	100	100
US citizen & perm. residents	4487	4532	4403	4492	4373	4269	4199	4417	4459	4567	4714	4685	4853	5378	5561	5579
%	84.9	84.5	83.7	83.0	80.6	79.6	77.8	76.1	75.5	73.5	72.4	70.1	70.1	74.9	76.1	72.5
US citizens–native born	4142	4188	4076	4128	4034	3894	3790	3953	4042	4087	4133	4113	4202	4247	4266	4285
%	78.3	78.1	77.4	76.2	74.3	72.6	70.2	68.1	68.4	65.8	63.5	61.6	60.7	59.1	58.3	55.7
US citizen–naturalized	140	164	143	174	150	174	169	160	162	194	250	215	236	260	240	286
%	2.6	3.1	2.7	3.2	2.8	3.2	3.1	2.8	2.7	3.1	3.8	3.2	3.4	3.6	3.3	3.7
US citizens	4282	4352	4219	4302	4184	4068	3959	4113	4204	4281	4383	4328	4438	4507	4506	4571
%	81.0	81.2	80.2	79.5	77.1	75.9	73.3	70.8	71.2	68.9	67.3	64.8	64.1	62.8	61.6	59.4
Permanent residents	205	180	184	190	189	201	240	304	255	286	331	357	415	871	1055	1008
%	3.9	3.4	3.5	3.5	3.5	3.8	4.4	5.2	4.3	4.6	5.1	5.3	6.0	12.1	14.4	13.1
Temp visas staying in US	156	176	191	223	263	264	333	413	476	606	928	1132	1165	960	910	1196
%	3.0	3.3	3.6	4.1	4.8	4.9	6.2	7.1	8.1	9.8	14.3	16.9	16.8	13.4	12.4	15.5
Temp visas leaving US	358	353	356	344	408	353	352	376	372	571	597	633	658	658	624	650
%	6.8	6.6	6.8	6.4	7.5	6.6	6.5	6.5	6.3	9.2	9.2	9.5	9.5	9.2	8.5	8.4
Total temporary residents	663	674	719	749	865	806	887	1032	1102	1472	1683	1886	1927	1726	1633	1939
%	12.5	12.6	13.7	13.8	15.9	15.0	16.4	17.8	18.7	23.7	25.9	28.2	27.8	24.0	22.3	25.2

Appendix E

Table E.5 (cont'd)

	1981	1982	1983	1984	1985	1986	1987	1988	1989	1990	1991	1992	1993	1994	1995	1996
Non responses	138	156	141	173	190	285	313	358	347	172	111	111	144	78	118	178
%	2.6	2.9	2.7	3.2	3.5	5.3	5.8	6.2	5.9	2.8	1.7	1.7	2.1	1.1	1.6	2.3

Appendix E

Table E.7 Number of US Life-Science Postdoctoral Fellows by Field and Citizenship, 1972-1995

Year	Biological	Agricultural	Total	Citizen	Noncitizen
1972	4845	303	5148		
1973	5237	242	5479		
1974	5231	272	5503		
1975	5785	273	6058		
1976	6282	349	6631	4831	1800
1977	6588	297	6885	5104	1781
1978	6848	227	7075	5134	1941
1979					
1980	7067	258	7325	5134	2191
1981	7668	292	7960	5477	2483
1982	7705	293	7998	5500	2498
1983	8324	308	8632	6000	2632
1984	8677	375	9052	6167	2885
1985	9112	371	9483	6167	3316
1986	9683	412	10095	6332	3763
1987	10358	442	10800	6517	4283
1988	10667	464	11131	6411	4720
1989	11425	517	11942	6648	5294

Appendix E

Table E.8 US Life-Science PhDs Who Took Postdoctoral Training Time Spent in Postdoctoral Positions, as Reported in 1995.

Year of PhD >>	1991-92	1989-90	1987-88	1985-86	1983-84	1981-82	1979-80	1977-78	1975-76	1973-74	1971-72	1969-70
Total PhDs who took postdoctoral training	4632	4502	4553	4410	4015	4024	4056	3552	3165	2670	2596	2158
less than 2 years	1332	1096	1291	1252	1010	1057	1132	1275	1235	1022	1059	1076
% less than 2 years	29	24	28	28	25	26	28	36	39	38	41	50
2-4 years	3300	2022	1609	1698	1560	1666	1687	1479	1320	1144	1035	671
% 2-4 years	71	45	35	39	39	41	42	42	42	43	40	31
greater than 4 years	NA	1384	1653	1460	1445	1301	1237	798	610	504	502	411
% greater than 4 years	NA	31	36	33	36	32	30	22	19	19	19	19

Appendix E

Table E.10 Demographic and Other Characteristics of US Life-Science PhDs, by 10-Year Cohort.

	\multicolumn{5}{c}{FY 86-96 Doctorate Recipients}	\multicolumn{5}{c}{FY 76-85 Doctorate Recipients}	\multicolumn{5}{c}{FY 66-75 Doctorate Recipients}
	Total Ph.D.s n	Women %	Temp. Res. %	Planning Postdoc %	Median Time to Degree yrs	Total Ph.D.s n	Women %	Temp. Res. %	Planning Postdoc %	Median Time to Degree yrs	Total Ph.D.s n	Women %	Temp. Res. %	Planning Postdoc %	Median Time to Degree yrs**
All doctorates	70989	36	23	51	7.9	51184	25	13	45	6.8	42647	15	14	34	6.2
Women	25695	100	18	53	7.9	12850	100	9	51	7.0	6579	100	9	41	6.2
Men	45096	0	25	50	7.8	38334	0	15	43	6.8	36068	0	15	33	6.1
Minorities (US cit or perm.)	2781	40	0	47	8.0	1324	35	0	33	7.2	296*	23	0	28	7.2
Whites (US cit or perm.)	43581	39	0	55	7.7	37813	26	0	50	6.6	9818*	20	0	40	6.1
US citizens & perm research	52681	39	0	55	7.8	43051	26	0	49	6.8	35964	16	0	36	6.0
Temporary research	16093	28	100	46	8.0	6754	17	100	28	7.3	5861	10	100	28	6.7
Top 26 institutions	19436	40	16	59	7.7	14187	30	10	54	6.7	11603	20	14	43	6.0
Other institutions	51553	35	25	48	7.9	36997	23	15	42	6.9	31044	14	14	31	6.2

Appendix F

DATA TABLES FOR CHAPTER 3

Tables 1-7 in this appendix show the fraction (labeled with an "F") and number (labeled with an "N") of graduates in a particular cohort (for example, 1-2 years after receipt of the PhD) who hold various types of positions. The matrix example on the following page illustrates how these career-progress matrices are to be read. Note that tables include only persons who at the time of doctorate were US citizens or permanent residents.

Table F.8 shows number of life-sccience PhDs by sector.

Table F.1 Career Progression of Life-Science PhDs–US Citizens and Permanent Residents.

Table F.2 Career Progression of Life-Science PhDs–Female US Citizens and Permanent Residents.

Table F.3 Career Progression of Life-Science PhDs–Male US Citizens and Permanent Residents.

Table F.4 Career Progression of Life-Science PhDs from 26 High-Quality Institutions–US Citizens and Permanent Residents.

Table F.5 Career Progression of Life-Science PhDs from Other Institutions–US Citizens and Permanent Residents.

Table F.6 Career Progression of Nonbiomedical Life-Science PhDs–US Citizens and Permanent Residents.

Table F.7 Career Progression of Biomedical Life-Science PhDs–US Citizens and Permanent Residents

Table F.8 Number of Citizen and Permanent Resident Life-Science PhDs by Sector, 1973-1995

Format for Tables F.1-F.7

Career Progression Matrix for Life-Science PhDs
(US Citizens and Permanent Residents with PhDs from US Universities)

	Survey Year	
	1973 .... **1993**	

Panel A: Fraction of 1-2 Year Cohort | 1971 & 1972 PhDs | 1991 & 1992 PhDs

    Not in Full-time S&E Work Force
Unemployed and seeking position
Part-time employed
Working outside science and engineering

    Full-time employed in S&E field
Tenure-track faculty position @ PhD Institutions
Tenure-track faculty position @ Other Inst
Postdoc total appointments in any sector
Other academic position
Industry
Federal labs and other government
Other jobs including self-employed

    Research Involvement in full-time S&E position
Engaged primarily in research
Supported by federal grants/contracts
Supported by HHS, NSF, and/or USDA

Panel B: Fraction of 3-4 Year Cohort | 1969 & 1970 PhDs | 1989 & 1990 PhDs

    Not in Full-time S&E Work Force
Unemployed and seeking position
Part-time employed
Working outside science and engineering

    Full-time employed in S&E field

Panel C: Fraction of 5-6 Year Cohort | 1967 & 1968 PhDs | 1987 & 1988 PhDs

    Not in Full-time S&E Work Force
Unemployed and seeking position
Part-time employed
Working outside science and engineering

    Full-time employed in S&E field

Appenidx F

	Survey Year	
	1973 . . . . **1993**	
Panel D: Fraction of 7-8 Year Cohort	1965 & 1966 PhDs ↓	1985 & 1986 PhDs ↓
Not in Full-time S&E Work Force		
Unemployed and seeking position		
Part-time employed		
Working outside science and engineering		
Full-time employed in S&E field		
Panel E: Fraction of 9-10 Year Cohort	1963 & 1964 PhDs ↓	1983 & 1984 PhDs ↓
Not in Full-time S&E Work Force		
Unemployed and seeking position		
Part-time employed		
Working outside science and engineering		
Full-time employed in S&E field		

Table F.1F Career Progression of Life-Science PhDs–US Citizens and Permanent Residents

	\multicolumn{7}{c}{Survey Year}
	1973	1977	1981	1985	1989	1993	1995
Panel A: Fraction of 1-2 Year Cohort							
Not in Full-time S&E Work Force							
Unemployed and seeking position	0.02	0.02	0.02	0.03	0.01	0.01	0.02
Part-time employed	0.03	0.02	0.03	0.03	0.03	0.03	0.02
Working outside science and engineering	0.02	0.03	0.01	0.03	0.02	0.03	0.03
Full-time employed in S&E field							
Tenure-track faculty position @ PhD Institutions	0.25	0.19	0.17	0.11	0.12	0.10	0.09
Tenure-track faculty position @ Other Inst	0.15	0.09	0.07	0.06	0.04	0.06	0.05
Postdoc total appointments in any sector	0.21	0.37	0.43	0.42	0.48	0.45	0.53
Other academic position	0.08	0.04	0.07	0.09	0.07	0.11	0.11
Industry	0.08	0.10	0.11	0.11	0.12	0.12	0.08
Federal labs and other government	0.11	0.09	0.05	0.06	0.07	0.06	0.04
Other jobs including self-employed	0.05	0.04	0.04	0.05	0.04	0.03	0.03
Research Involvement in full-time S&E position							
Engaged primarily in research	0.54	0.59	0.69	0.67	0.76	0.71	0.70
Supported by federal grants/contracts	0.52	0.52	0.57	0.45	n/a	0.44	0.49
Supported by HHS, NSF, and/or USDA	0.44	0.42	0.48	0.36	n/a	0.39	0.45
Panel B: Fraction of 3-4 Year Cohort							
Not in Full-time S&E Work Force							
Unemployed and seeking position	0.01	0.03	0.02	0.02	0.01	0.01	0.02
Part-time employed	0.02	0.02	0.02	0.02	0.04	0.05	0.03
Working outside science and engineering	0.03	0.02	0.02	0.01	0.02	0.05	0.05
Full-time employed in S&E field							
Tenure-track faculty position @ PhD Institutions	0.40	0.31	0.29	0.24	0.21	0.17	0.13
Tenure-track faculty position @ Other Inst	0.15	0.13	0.08	0.06	0.03	0.07	0.07
Postdoc total appointments in any sector	0.06	0.11	0.19	0.21	0.22	0.28	0.29
Other academic position	0.06	0.06	0.08	0.09	0.12	0.10	0.13
Industry	0.11	0.13	0.16	0.21	0.20	0.17	0.16
Federal labs and other government	0.11	0.10	0.09	0.08	0.10	0.07	0.07
Other jobs including self-employed	0.06	0.08	0.05	0.06	0.06	0.04	0.04
Research Involvement in full-time S&E position							
Engaged primarily in research	0.44	0.45	0.64	0.61	0.70	0.68	0.63
Supported by federal grants/contracts	0.51	0.46	0.44	0.41	n/a	0.41	0.45

	Survey Year						
	1973	1977	1981	1985	1989	1993	1995

Panel C: Fraction of 5-6 Year Cohort

Not in Full-time S&E Work Force

Unemployed and seeking position	0.01	0.02	0.02	0.01	0.01	0.01	0.02
Part-time employed	0.03	0.03	0.02	0.04	0.01	0.04	0.05
Working outside science and engineering	0.02	0.04	0.02	0.04	0.03	0.04	0.06

Full-time employed in S&E field

Tenure-track faculty position @ PhD Institutions	0.42	0.38	0.34	0.32	0.27	0.27	0.22
Tenure-track faculty position @ Other Inst	0.15	0.15	0.10	0.07	0.08	0.06	0.08
Postdoc total appointments in any sector	0.02	0.05	0.07	0.10	0.10	0.12	0.14
Other academic position	0.03	0.04	0.07	0.06	0.10	0.08	0.12
Industry	0.11	0.11	0.16	0.20	0.24	0.22	0.19
Federal labs and other government	0.15	0.13	0.13	0.10	0.09	0.10	0.08
Other jobs including self-employed	0.06	0.07	0.08	0.06	0.07	0.06	0.05

Research Involvement in full-time S&E position

Engaged primarily in research	0.45	0.42	0.51	0.54	0.63	0.64	0.65
Supported by federal grants/contracts	0.46	0.46	0.43	0.36	n/a	0.37	0.40
Supported by HHS, NSF, and/or USDA	0.39	0.37	0.37	0.30	n/a	0.33	0.35

Panel D: Fraction of 7-8 Year Cohort

Not in Full-time S&E Work Force

Unemployed and seeking position	0.01	0.01	0.01	0.01	0.00	0.01	0.01
Part-time employed	0.02	0.02	0.02	0.02	0.03	0.04	0.03
Working outside science and engineering	0.03	0.03	0.02	0.04	0.05	0.06	0.05

Full-time employed in S&E field

Tenure-track faculty position @ PhD Institutions	0.47	0.44	0.40	0.38	0.32	0.31	0.32
Tenure-track faculty position @ Other Inst	0.15	0.14	0.11	0.08	0.05	0.05	0.07
Postdoc total appointments in any sector	0.01	0.03	0.03	0.02	0.05	0.06	0.06
Other academic position	0.02	0.03	0.05	0.06	0.09	0.10	0.08
Industry	0.11	0.11	0.17	0.20	0.24	0.21	0.22
Federal labs and other government	0.13	0.13	0.11	0.11	0.08	0.11	0.10
Other jobs including self-employed	0.05	0.07	0.09	0.08	0.07	0.06	0.07

Research Involvement in full-time S&E position

Engaged primarily in research	0.42	0.41	0.43	0.51	0.61	0.61	0.61
Supported by federal grants/contracts	0.51	0.45	0.39	0.38	n/a	0.39	0.39
Supported by HHS, NSF, and/or USDA	0.43	0.37	0.31	0.32	n/a	0.35	0.35

	Survey Year						
	1973	1977	1981	1985	1989	1993	1995
Panel E: Fraction of 9-10 Year Cohort							
Not in Full-time S&E Work Force							
Unemployed and seeking position	0.01	0.01	0.02	0.01	0.00	0.01	0.01
Part-time employed	0.02	0.02	0.02	0.03	0.03	0.02	0.03
Working outside science and engineering	0.03	0.04	0.04	0.04	0.04	0.07	0.04
Full-time employed in S&E field							
Tenure-track faculty position @ PhD Institutions	0.47	0.42	0.39	0.35	0.36	0.31	0.34
Tenure-track faculty position @ Other Inst	0.14	0.15	0.15	0.10	0.06	0.09	0.05
Postdoc total appointments in any sector	0.00	0.01	0.01	0.02	0.02	0.03	0.03
Other academic position	0.01	0.01	0.05	0.06	0.06	0.07	0.09
Industry	0.12	0.11	0.11	0.18	0.23	0.22	0.23
Federal labs and other government	0.14	0.16	0.16	0.13	0.11	0.11	0.11
Other jobs including self-employed	0.05	0.05	0.06	0.09	0.08	0.07	0.07
Research Involvement in full-time S&E position							
Engaged primarily in research	0.41	0.43	0.40	0.44	0.54	0.57	0.58
Supported by federal grants/contracts	0.44	0.43	0.36	0.37	n/a	0.37	0.38
Supported by HHS, NSF, and/or USDA	0.38	0.34	0.30	0.31	n/a	0.33	0.36

Table F.1N Career Progresson of Life-Science PhDs–US Citizens and Permanent Residents

	\multicolumn{7}{c}{Survey Year}
	1973	1977	1981	1985	1989	1993	1995
Panel A: Number in 1-2 Year Cohort							
Not in Full-time S&E Work Force							
Unemployed and seeking position	144	175	176	220	111	95	211
Part-time employed	195	128	222	236	300	221	202
Working outside science and engineering	136	220	78	262	166	291	319
Full-time employed in S&E field							
Tenure-track faculty position @ PhD Institutions	1936	1473	1347	956	1002	878	853
Tenure-track faculty position @ Other Inst	1158	683	530	510	315	486	469
Postdoc total appointments in any sector	1659	2847	3447	3498	4141	3952	4980
Other academic position	612	322	539	771	574	968	1031
Industry	659	783	906	955	1013	1075	745
Federal labs and other government	864	683	430	484	597	551	359
Other jobs including self-employed	412	321	358	438	381	288	293
Research Involvement in full-time S&E position							
Engaged primarily in research	3963	4214	5197	5132	6093	5799	6131
Supported by federal grants/contracts	3807	3719	4313	3433	n/a	3645	4308
Supported by HHS, NSF, and/or USDA	3188	3013	3656	2727	n/a	3179	3899
Panel B: Number in 3-4 Year Cohort							
Not in Full-time S&E Work Force							
Unemployed and seeking position	70	210	127	132	75	64	173
Part-time employed	152	113	123	205	295	394	288
Working outside science and engineering	179	185	157	94	160	387	458
Full-time employed in S&E field							
Tenure-track faculty position @ PhD Institutions	2684	2336	2301	2017	1694	1392	1162
Tenure-track faculty position @ Other Inst	1009	983	631	491	247	543	657
Postdoc total appointments in any sector	384	856	1456	1742	1830	2352	2564
Other academic position	379	418	643	773	974	871	1106
Industry	706	1007	1289	1804	1593	1419	1432
Federal labs and other government	758	765	736	654	802	552	656
Other jobs including self-employed	390	605	383	521	487	355	313
Research Involvement in full-time S&E position							
Engaged primarily in research	2776	3124	4758	4861	5319	5059	4995
Supported by federal grants/contracts	3223	3228	3304	3310	n/a	3046	3551
Supported by HHS, NSF, and/or USDA	2680	2656	2743	2719	n/a	2674	3141

	Survey Year						
	1973	1977	1981	1985	1989	1993	1995

Panel C: Number in 5-6 Year Cohort

Not in Full-time S&E Work Force

Unemployed and seeking position	43	132	152	54	92	105	166
Part-time employed	166	217	149	301	117	308	398
Working outside science and engineering	102	296	172	289	246	341	460

Full-time employed in S&E field

Tenure-track faculty position @ PhD Institutions	2211	3088	2591	2573	2245	2167	1816
Tenure-track faculty position @ Other Inst	776	1204	783	605	651	476	637
Postdoc total appointments in any sector	94	383	516	815	839	1001	1171
Other academic position	177	307	538	528	861	621	978
Industry	549	905	1214	1657	1974	1820	1572
Federal labs and other government	805	1065	962	800	734	778	633
Other jobs including self-employed	300	613	579	517	563	505	396

Research Involvement in full-time S&E position

Engaged primarily in research	2186	3200	3676	4015	4991	4702	4683
Supported by federal grants/contracts	2271	3514	3067	2688	n/a	2736	2850
Supported by HHS, NSF, and/or USDA	1926	2814	2625	2271	n/a	2454	2512

Panel D: Number in 7-8 Year Cohort

Not in Full-time S&E Work Force

Unemployed and seeking position	36	95	69	94	32	66	113
Part-time employed	75	119	169	166	300	308	227
Working outside science and engineering	121	199	167	321	429	463	365

Full-time employed in S&E field

Tenure-track faculty position @ PhD Institutions	1896	2996	2899	2925	2863	2493	2521
Tenure-track faculty position @ Other Inst	581	930	798	631	488	445	568
Postdoc total appointments in any sector	45	191	185	156	437	460	465
Other academic position	77	183	333	491	829	815	594
Industry	451	772	1258	1583	2125	1737	1734
Federal labs and other government	514	870	795	829	730	870	770
Other jobs including self-employed	201	471	649	593	653	504	557

Research Involvement in full-time S&E position

Engaged primarily in research	1571	2640	2949	3697	4958	4491	4401
Supported by federal grants/contracts	1916	2908	2670	2716	n/a	2862	2826
Supported by HHS, NSF, and/or USDA	1615	2401	2145	2338	n/a	2571	2557

Appenidx F

	Survey Year						
	1973	1977	1981	1985	1989	1993	1995

Panel E: Number in 9-10 Year Cohort

Not in Full-time S&E Work Force

Unemployed and seeking position	30	77	128	81	41	111	44
Part-time employed	72	129	180	232	277	184	242
Working outside science and engineering	115	224	286	297	296	559	340

Full-time employed in S&E field

Tenure-track faculty position @ PhD Institutions	1672	2267	3056	2565	3005	2588	2725
Tenure-track faculty position @ Other Inst	501	791	1156	743	508	729	422
Postdoc total appointments in any sector	7	64	59	132	171	264	228
Other academic position	47	63	397	425	491	574	745
Industry	430	617	847	1321	1884	1861	1882
Federal labs and other government	485	877	1260	939	895	964	882
Other jobs including self-employed	162	277	472	676	691	564	588

Research Involvement in full-time S&E position

Engaged primarily in research	1357	2144	2903	3004	4092	4334	4345
Supported by federal grants/contracts	1446	2142	2601	2484	n/a	2812	2868
Supported by HHS, NSF, and/or USDA	1266	1699	2207	2082	n/a	2496	2664

Table F.2F Career Progression of Life-Science PhDs—Female US Citizens and Permanenet Residents

	\multicolumn{7}{c}{Survey Year}
	1973	1977	1981	1985	1989	1993	1995
Panel A: Fraction of 1-2 Year Cohort							
Not in Full-time S&E Work Force							
Unemployed and seeking position	0.05	0.05	0.03	0.03	0.02	0.01	0.03
Part-time employed	0.09	0.05	0.06	0.05	0.07	0.04	0.03
Working outside science and engineering	0.03	0.04	0.01	0.05	0.02	0.04	0.04
Full-time employed in S&E field							
Tenure-track faculty position @ PhD Institutions	0.15	0.11	0.10	0.09	0.12	0.09	0.07
Tenure-track faculty position @ Other Inst	0.12	0.13	0.05	0.05	0.03	0.05	0.04
Postdoc total appointments in any sector	0.30	0.45	0.55	0.47	0.48	0.47	0.55
Other academic position	0.12	0.04	0.07	0.06	0.09	0.11	0.10
Industry	0.04	0.04	0.08	0.09	0.07	0.10	0.08
Federal labs and other government	0.04	0.04	0.03	0.05	0.06	0.05	0.03
Other jobs including self-employed	0.06	0.05	0.03	0.05	0.03	0.03	0.03
Research Involvement in full-time S&E position							
11 Engaged primarily in research	0.68	0.63	0.78	0.68	0.76	0.72	0.72
12 Supported by federal grants/contracts	0.62	0.59	0.60	0.50	n/a	0.46	0.55
13 Supported by HHS, NSF, and/or USDA	0.53	0.52	0.56	0.43	n/a	0.40	0.50
Panel B: Fraction of 3-4 Year Cohort							
Not in Full-time S&E Work Force							
Unemployed and seeking position	0.04	0.03	0.04	0.03	0.03	0.02	0.02
Part-time employed	0.11	0.06	0.07	0.06	0.04	0.09	0.07
Working outside science and engineering	0.04	0.02	0.03	0.03	0.04	0.05	0.05
Full-time employed in S&E field							
Tenure-track faculty position @ PhD Institutions	0.27	0.28	0.21	0.19	0.18	0.16	0.10
Tenure-track faculty position @ Other Inst	0.17	0.16	0.09	0.09	0.03	0.08	0.09
Postdoc total appointments in any sector	0.10	0.19	0.25	0.24	0.23	0.28	0.32
Other academic position	0.12	0.06	0.06	0.12	0.15	0.10	0.11
Industry	0.03	0.04	0.09	0.13	0.14	0.13	0.12
Federal labs and other government	0.05	0.06	0.09	0.07	0.07	0.05	0.07
Other jobs including self-employed	0.08	0.09	0.07	0.05	0.08	0.04	0.05
Research Involvement in full-time S&E position							
Engaged primarily in research	0.49	0.52	0.59	0.60	0.72	0.64	0.61
Supported by federal grants/contracts	0.58	0.55	0.49	0.45	n/a	0.42	0.46
Supported by HHS, NSF, and/or USDA	0.46	0.47	0.41	0.40	n/a	0.38	0.42

	Survey Year						
	1973	1977	1981	1985	1989	1993	1995

Panel C: Fraction of 5-6 Year Cohort

Not in Full-time S&E Work Force

Unemployed and seeking position	0.04	0.03	0.03	0.01	0.03	0.02	0.01
Part-time employed	0.18	0.12	0.06	0.08	0.04	0.09	0.09
Working outside science and engineering	0.05	0.02	0.03	0.04	0.03	0.04	0.07

Full-time employed in S&E field

Tenure-track faculty position @ PhD Institutions	0.33	0.31	0.28	0.26	0.24	0.23	0.19
Tenure-track faculty position @ Other Inst	0.13	0.15	0.14	0.05	0.08	0.07	0.10
Postdoc total appointments in any sector	0.07	0.09	0.13	0.10	0.09	0.13	0.11
Other academic position	0.05	0.06	0.08	0.14	0.10	0.09	0.18
Industry	0.04	0.07	0.06	0.15	0.20	0.18	0.17
Federal labs and other government	0.05	0.07	0.09	0.10	0.10	0.10	0.06
Other jobs including self-employed	0.07	0.08	0.10	0.06	0.09	0.06	0.03

Research Involvement in full-time S&E position

Engaged primarily in research	0.53	0.49	0.54	0.62	0.60	0.65	0.61
Supported by federal grants/contracts	0.58	0.48	0.48	0.39	n/a	0.40	0.44
Supported by HHS, NSF, and/or USDA	0.51	0.36	0.42	0.36	n/a	0.38	0.38

Panel D: Fraction of 7-8 Year Cohort

Not in Full-time S&E Work Force

Unemployed and seeking position	0.05	0.03	0.04	0.03	0.01	0.01	0.03
Part-time employed	0.15	0.09	0.05	0.08	0.09	0.11	0.07
Working outside science and engineering	0.05	0.01	0.05	0.03	0.06	0.05	0.03

Full-time employed in S&E field

Tenure-track faculty position @ PhD Institutions	0.29	0.38	0.39	0.29	0.31	0.29	0.29
Tenure-track faculty position @ Other Inst	0.18	0.17	0.14	0.11	0.06	0.07	0.09
Postdoc total appointments in any sector	0.06	0.08	0.03	0.04	0.05	0.05	0.06
Other academic position	0.06	0.04	0.07	0.11	0.11	0.12	0.08
Industry	0.03	0.04	0.04	0.13	0.19	0.15	0.19
Federal labs and other government	0.09	0.07	0.07	0.09	0.06	0.10	0.08
Other jobs including self-employed	0.04	0.09	0.11	0.08	0.06	0.06	0.07

Research Involvement in full-time S&E position

Engaged primarily in research	0.47	0.43	0.51	0.52	0.63	0.6	0.58
Supported by federal grants/contracts	0.50	0.54	0.45	0.46	n/a	0.45	0.39
Supported by HHS, NSF, and/or USDA	0.40	0.45	0.39	0.37	n/a	0.41	0.36

	Survey Year						
	1973	1977	1981	1985	1989	1993	1995
Panel E: Fraction of 9-10 Year Cohort							
Not in Full-time S&E Work Force							
Unemployed and seeking position	0.06	0.04	0.06	0.01	0.02	0.01	0.01
Part-time employed	0.11	0.15	0.07	0.10	0.12	0.06	0.08
Working outside science and engineering	0.04	0.04	0.02	0.03	0.01	0.09	0.05
Full-time employed in S&E field							
Tenure-track faculty position @ PhD Institutions	0.32	0.39	0.35	0.32	0.34	0.29	0.36
Tenure-track faculty position @ Other Inst	0.22	0.13	0.14	0.13	0.06	0.09	0.06
Postdoc total appointments in any sector	0.02	0.04	0.02	0.05	0.02	0.03	0.01
Other academic position	0.05	0.02	0.08	0.08	0.07	0.09	0.11
Industry	0.05	0.05	0.10	0.09	0.20	0.18	0.15
Federal labs and other government	0.07	0.05	0.08	0.09	0.08	0.11	0.10
Other jobs including self-employed	0.06	0.08	0.09	0.10	0.07	0.04	0.07
Research Involvement in full-time S&E position							
Engaged primarily in research	0.42	0.44	0.45	0.47	0.52	0.60	0.57
Supported by federal grants/contracts	0.50	0.50	0.43	0.41	n/a	0.39	0.45
Supported by HHS, NSF, and/or USDA	0.43	0.43	0.39	0.38	n/a	0.35	0.44

Table F.2N Career Progression of Life-Science PhDs–Female US Citizens and Permanent Residents

	\multicolumn{7}{c}{Survey Year}
	1973	1977	1981	1985	1989	1993	1995
Panel A: Number in 1-2 Year Cohort							
Not in Full-time S&E Work Force							
Unemployed and seeking position	59	72	59	75	54	39	122
Part-time employed	105	72	122	132	197	153	102
Working outside science and engineering	29	57	19	115	62	155	141
Full-time employed in S&E field							
Tenure-track faculty position @ PhD Institutions	172	173	194	233	370	314	288
Tenure-track faculty position @ Other Inst	130	214	95	139	94	186	172
Postdoc total appointments in any sector	338	725	1103	1199	1437	1620	2171
Other academic position	133	71	138	164	260	379	389
Industry	43	69	166	229	218	340	295
Federal labs and other government	45	68	61	131	183	174	126
Other jobs including self-employed	71	74	54	128	92	113	109
Research Involvement in full-time S&E position							
Engaged primarily in research	638	879	1409	1521	2008	2245	2571
Supported by federal grants/contracts	580	818	1092	1110	n/a	1437	1944
Supported by HHS, NSF, and/or USDA	492	720	1006	966	n/a	1239	1768
Panel B: Number in 3-4 Year Cohort							
Not in Full-time S&E Work Force							
Unemployed and seeking position	34	43	63	72	71	54	59
Part-time employed	95	77	116	125	110	264	233
Working outside science and engineering	31	31	48	57	105	165	174
Full-time employed in S&E field							
Tenure-track faculty position @ PhD Institutions	224	356	336	423	451	487	346
Tenure-track faculty position @ Other Inst	139	205	144	204	84	253	295
Postdoc total appointments in any sector	80	241	394	536	571	876	1116
Other academic position	100	82	96	281	372	311	374
Industry	21	50	136	301	350	394	406
Federal labs and other government	42	82	145	154	167	164	248
Other jobs including self-employed	66	122	113	106	188	129	183
Research Involvement in full-time S&E position							
Engaged primarily in research	327	589	803	1193	1569	1668	1808
Supported by federal grants/contracts	387	626	671	905	n/a	1106	1370
Supported by HHS, NSF, and/or USDA	309	531	560	793	n/a	988	1251

	Survey Year						
	1973	1977	1981	1985	1989	1993	1995

Panel C: Number in 5-6 Year Cohort

Not in Full-time S&E Work Force

Unemployed and seeking position	24	36	42	18	67	49	39
Part-time employed	116	144	100	154	113	253	268
Working outside science and engineering	31	28	43	79	78	100	201

Full-time employed in S&E field

Tenure-track faculty position @ PhD Institutions	217	376	439	515	603	646	567
Tenure-track faculty position @ Other Inst	88	183	217	97	208	190	306
Postdoc total appointments in any sector	43	106	206	204	228	369	320
Other academic position	34	71	126	273	264	240	540
Industry	25	91	89	296	523	514	514
Federal labs and other government	34	85	133	192	249	291	190
Other jobs including self-employed	47	98	156	123	232	161	81

Research Involvement in full-time S&E position

Engaged primarily in research	257	495	732	1046	1395	1565	1526
Supported by federal grants/contracts	284	482	661	667	n/a	953	1099
Supported by HHS, NSF, and/or USDA	249	362	568	620	n/a	914	966

Panel D: Number in 7-8 Year Cohort

Not in Full-time S&E Work Force

Unemployed and seeking position	21	24	46	45	16	28	89
Part-time employed	66	76	66	117	227	272	195
Working outside science and engineering	21	11	58	51	138	119	94

Full-time employed in S&E field

Tenure-track faculty position @ PhD Institutions	127	320	490	444	749	740	769
Tenure-track faculty position @ Other Inst	81	142	180	173	146	174	248
Postdoc total appointments in any sector	25	67	44	63	119	137	166
Other academic position	27	38	94	164	261	318	224
Industry	15	34	53	200	449	376	517
Federal labs and other government	38	63	94	138	144	260	217
Other jobs including self-employed	17	72	134	128	142	164	177

Research Involvement in full-time S&E position

Engaged primarily in research	155	315	559	680	1263	1305	1339
Supported by federal grants/contracts	164	401	494	609	n/a	980	897
Supported by HHS, NSF, and/or USDA	132	333	420	482	n/a	879	843

	Survey Year						
	1973	1977	1981	1985	1989	1993	1995
Panel E: Number in 9-10 Year Cohort							
Not in Full-time S&E Work Force							
Unemployed and seeking position	21	28	64	20	41	31	19
Part-time employed	39	103	76	147	238	153	199
Working outside science and engineering	15	28	26	45	25	231	121
Full-time employed in S&E field							
Tenure-track faculty position @ PhD Institutions	111	274	395	478	673	729	935
Tenure-track faculty position @ Other Inst	75	89	163	188	125	226	163
Postdoc total appointments in any sector	7	31	19	71	38	64	32
Other academic position	16	16	88	119	145	220	275
Industry	17	37	113	140	403	456	396
Federal labs and other government	25	38	87	136	148	271	259
Other jobs including self-employed	21	56	103	154	130	111	175
Research Involvement in full-time S&E position							
Engaged primarily in research	115	237	433	598	867	1247	1285
Supported by federal grants/contracts	135	271	421	528	n/a	812	1010
Supported by HHS, NSF, and/or USDA	116	230	380	483	n/a	718	984

Table F.3F Career Progression of Life-Science PhDs–Male US Citizens and Permanent Residents.

	\multicolumn{7}{c}{Survey Year}
	1973	1977	1981	1985	1989	1993	1995
Panel A: Fraction of 1-2 Year Cohort							
Not in Full-time S&E Work Force							
Unemployed and seeking position	0.01	0.02	0.02	0.03	0.01	0.01	0.02
Part-time employed	0.01	0.01	0.02	0.02	0.02	0.01	0.02
Working outside science and engineering	0.02	0.03	0.01	0.03	0.02	0.03	0.03
Full-time employed in S&E field							
Tenure-track faculty position @ PhD Institutions	0.27	0.22	0.19	0.12	0.11	0.11	0.10
Tenure-track faculty position @ Other Inst	0.15	0.08	0.07	0.06	0.04	0.06	0.05
Postdoc total appointments (in any sector)	0.20	0.35	0.39	0.40	0.48	0.44	0.51
Other academic position	0.07	0.04	0.07	0.10	0.06	0.11	0.12
Industry	0.09	0.12	0.12	0.13	0.14	0.14	0.08
Federal labs and other government	0.12	0.10	0.06	0.06	0.07	0.07	0.04
Other jobs (including self-employed)	0.05	0.04	0.05	0.05	0.05	0.03	0.03
Research Involvement (in full-time S&E position)							
Engaged primarily in research	0.52	0.58	0.66	0.67	0.76	0.70	0.69
Supported by federal grants/contracts	0.51	0.51	0.56	0.43	n/a	0.44	0.46
Supported by HHS, NSF, and/or USDA	0.42	0.40	0.46	0.33	n/a	0.38	0.41
Panel B: Fraction of 3-4 Year Cohort							
Not in Full-time S&E Work Force							
Unemployed and seeking position	0.01	0.03	0.01	0.01	0.00	0.00	0.02
Part-time employed	0.01	0.01	0.00	0.01	0.03	0.02	0.01
Working outside science and engineering	0.03	0.02	0.02	0.01	0.01	0.04	0.05
Full-time employed in S&E field							
Tenure-track faculty position @ PhD Institutions	0.42	0.32	0.31	0.26	0.22	0.17	0.15
Tenure-track faculty position @ Other Inst	0.15	0.13	0.08	0.05	0.03	0.06	0.07
Postdoc total appointments (in any sector)	0.05	0.10	0.17	0.20	0.22	0.28	0.27
Other academic position	0.05	0.05	0.09	0.08	0.11	0.11	0.14
Industry	0.12	0.15	0.18	0.24	0.22	0.20	0.19
Federal labs and other government	0.12	0.11	0.09	0.08	0.11	0.07	0.08
Other jobs (including self-employed)	0.06	0.08	0.04	0.07	0.05	0.04	0.02
Research Involvement (in full-time S&E position)							
Engaged primarily in research	0.43	0.43	0.65	0.61	0.69	0.70	0.65
Supported by federal grants/contracts	0.50	0.45	0.43	0.40	n/a	0.40	0.44
Supported by HHS, NSF, and/or USDA	0.42	0.36	0.36	0.32	n/a	0.35	0.38

Appenidx F

	Survey Year						
	1973	1977	1981	1985	1989	1993	1995

Panel C: Fraction of 5-6 Year Cohort

Not in Full-time S&E Work Force

Unemployed and seeking position	0.00	0.01	0.02	0.01	0.00	0.01	0.02
Part-time employed	0.01	0.01	0.01	0.02	0.00	0.01	0.02
Working outside science and engineering	0.02	0.04	0.02	0.03	0.03	0.05	0.05

Full-time employed in S&E field

Tenure-track faculty position @ PhD Institutions	0.44	0.39	0.35	0.33	0.29	0.29	0.24
Tenure-track faculty position @ Other Inst	0.15	0.15	0.09	0.08	0.08	0.05	0.06
Postdoc total appointments (in any sector)	0.01	0.04	0.05	0.10	0.11	0.12	0.16
Other academic position	0.03	0.03	0.07	0.04	0.10	0.07	0.08
Industry	0.11	0.12	0.18	0.22	0.25	0.25	0.20
Federal labs and other government	0.17	0.14	0.14	0.10	0.08	0.09	0.09
Other jobs (including self-employed)	0.06	0.07	0.07	0.06	0.06	0.06	0.06

Research Involvement (in full-time S&E position)

Engaged primarily in research	0.44	0.41	0.51	0.51	0.65	0.63	0.67
Supported by federal grants/contracts	0.45	0.46	0.41	0.35	n/a	0.36	0.37
Supported by HHS, NSF, and/or USDA	0.38	0.37	0.35	0.28	n/a	0.31	0.33

Panel D: Fraction of 7-8 Year Cohort

Not in Full-time S&E Work Force

Unemployed and seeking position	0.00	0.01	0.00	0.01	0.00	0.01	0.00
Part-time employed	0.00	0.01	0.02	0.01	0.01	0.01	0.01
Working outside science and engineering	0.03	0.03	0.02	0.04	0.04	0.06	0.05

Full-time employed in S&E field

Tenure-track faculty position @ PhD Institutions	0.50	0.45	0.40	0.40	0.33	0.31	0.34
Tenure-track faculty position @ Other Inst	0.14	0.13	0.10	0.07	0.05	0.05	0.06
Postdoc total appointments (in any sector)	0.01	0.02	0.02	0.01	0.05	0.06	0.06
Other academic position	0.01	0.02	0.04	0.05	0.09	0.09	0.07
Industry	0.12	0.12	0.20	0.22	0.26	0.24	0.23
Federal labs and other government	0.13	0.13	0.12	0.11	0.09	0.11	0.11
Other jobs (including self-employed)	0.05	0.07	0.08	0.07	0.08	0.06	0.07

Research Involvement (in full-time S&E position)

Engaged primarily in research	0.41	0.41	0.41	0.51	0.60	0.62	0.63
Supported by federal grants/contracts	0.51	0.44	0.37	0.36	n/a	0.37	0.39
Supported by HHS, NSF, and/or USDA	0.43	0.36	0.30	0.31	n/a	0.33	0.35

	Survey Year						
	1973	1977	1981	1985	1989	1993	1995
Panel E: Fraction of 9-10 Year Cohort							
Not in Full-time S&E Work Force							
Unemployed and seeking position	0.00	0.01	0.01	0.01	0.00	0.01	0.00
Part-time employed	0.01	0.01	0.02	0.01	0.01	0.01	0.01
Working outside science and engineering	0.03	0.04	0.04	0.04	0.04	0.06	0.04
Full-time employed in S&E field							
Tenure-track faculty position @ PhD Institutions	0.49	0.43	0.40	0.35	0.37	0.31	0.32
Tenure-track faculty position @ Other Inst	0.13	0.15	0.15	0.09	0.06	0.09	0.05
Postdoc total appointments (in any sector)	0.00	0.01	0.01	0.01	0.02	0.03	0.04
Other academic position	0.01	0.01	0.05	0.05	0.05	0.06	0.09
Industry	0.13	0.12	0.11	0.20	0.24	0.24	0.27
Federal labs and other government	0.14	0.18	0.17	0.14	0.12	0.12	0.11
Other jobs (including self-employed)	0.04	0.05	0.06	0.09	0.09	0.08	0.07
Research Involvement (in full-time S&E position)							
Engaged primarily in research	0.41	0.43	0.39	0.44	0.54	0.56	0.58
Supported by federal grants/contracts	0.43	0.42	0.35	0.35	n/a	0.37	0.35
Supported by HHS, NSF, and/or USDA	0.38	0.33	0.29	0.29	n/a	0.33	0.32

Table F.3N Career Progression of Life-Science PhDs–Male US Citizens and Permanent Residents

	Survey Year						
	1973	1977	1981	1985	1989	1993	1995
Panel A: Number in 1-2 Year Cohort							
Not in Full-time S&E Work Force							
Unemployed and seeking position	85	103	117	145	57	56	89
Part-time employed	90	56	100	104	103	68	100
Working outside science and engineering	107	163	59	147	104	136	178
Full-time employed in S&E field							
Tenure-track faculty position @ PhD Institutions	1764	1300	1153	723	632	564	565
Tenure-track faculty position @ Other Inst	1028	469	435	371	221	300	297
Postdoc total appointments (in any sector)	1321	2122	2344	2299	2704	2332	2809
Other academic position	479	251	401	607	314	589	642
Industry	616	714	740	726	795	735	450
Federal labs and other government	819	615	369	353	414	377	233
Other jobs (including self-employed)	341	247	304	310	289	175	184
Research Involvement (in full-time S&E position)							
Engaged primarily in research	3325	3335	3788	3611	4085	3554	3560
Supported by federal grants/contracts	3227	2901	3221	2323	n/a	2208	2364
Supported by HHS, NSF, and/or USDA	2696	2293	2650	1761	n/a	1940	2131
Panel B: Number in 3-4 Year Cohort							
Not in Full-time S&E Work Force							
Unemployed and seeking position	36	167	64	60	4	10	114
Part-time employed	57	36	7	80	185	130	55
Working outside science and engineering	148	154	109	37	55	222	284
Full-time employed in S&E field							
Tenure-track faculty position @ PhD Institutions	2460	1980	1965	1594	1243	905	816
Tenure-track faculty position @ Other Inst	870	778	487	287	163	289	362
Postdoc total appointments (in any sector)	304	615	1062	1206	1259	1476	1448
Other academic position	279	336	547	492	602	560	732
Industry	685	957	1153	1503	1243	1025	1026
Federal labs and other government	716	683	591	500	635	388	408
Other jobs (including self-employed)	324	483	270	415	299	226	130
Research Involvement (in full-time S&E position)							
Engaged primarily in research	2449	2535	3955	3668	3750	3391	3187
Supported by federal grants/contracts	2836	2602	2633	2405	n/a	1940	2181
Supported by HHS, NSF, and/or USDA	2371	2125	2183	1926	n/a	1686	1890

	Survey Year						
	1973	1977	1981	1985	1989	1993	1995
Panel C: Number in 5-6 Year Cohort							
Not in Full-time S&E Work Force							
Unemployed and seeking position	19	96	110	36	25	56	127
Part-time employed	50	73	49	147	4	55	130
Working outside science and engineering	71	268	129	210	168	241	259
Full-time employed in S&E field							
Tenure-track faculty position @ PhD Institutions	1994	2712	2152	2058	1642	1521	1249
Tenure-track faculty position @ Other Inst	688	1021	566	508	443	286	331
Postdoc total appointments (in any sector)	51	277	310	611	611	632	851
Other academic position	143	236	412	255	597	381	438
Industry	524	814	1125	1361	1451	1306	1058
Federal labs and other government	771	980	829	608	485	487	443
Other jobs (including self-employed)	253	515	423	394	331	344	315
Research Involvement (in full-time S&E position)							
Engaged primarily in research	1929	2705	2944	2969	3596	3137	3157
Supported by federal grants/contracts	1987	3032	2406	2021	n/a	1783	1751
Supported by HHS, NSF, and/or USDA	1677	2452	2057	1651	n/a	1540	1546
Panel D: Number in 7-8 Year Cohort							
Not in Full-time S&E Work Force							
Unemployed and seeking position	15	71	23	49	16	38	24
Part-time employed	9	43	103	49	73	36	32
Working outside science and engineering	100	188	109	270	291	344	271
Full-time employed in S&E field							
Tenure-track faculty position @ PhD Institutions	1769	2676	2409	2481	2114	1753	1752
Tenure-track faculty position @ Other Inst	500	788	618	458	342	271	320
Postdoc total appointments (in any sector)	20	124	141	93	318	323	299
Other academic position	50	145	239	327	568	497	370
Industry	436	738	1205	1383	1676	1361	1217
Federal labs and other government	476	807	701	691	586	610	553
Other jobs (including self-employed)	184	399	515	465	511	340	380
Research Involvement (in full-time S&E position)							
Engaged primarily in research	1416	2325	2390	3017	3695	3186	3062
Supported by federal grants/contracts	1752	2507	2176	2107	n/a	1882	1929
Supported by HHS, NSF, and/or USDA	1483	2068	1725	1856	n/a	1692	1714

Appenidx F

	Survey Year						
	1973	1977	1981	1985	1989	1993	1995

Panel E: Number in 9-10 Year Cohort

Not in Full-time S&E Work Force
Unemployed and seeking position	9	49	64	61		80	25
Part-time employed	33	26	104	85	39	31	43
Working outside science and engineering	100	196	260	252	271	328	219

Full-time employed in S&E field
Tenure-track faculty position @ PhD Institutions	1561	1993	2661	2087	2332	1859	1790
Tenure-track faculty position @ Other Inst	426	702	993	555	383	503	259
Postdoc total appointments (in any sector)	.	33	40	61	133	200	196
Other academic position	31	47	309	306	346	354	470
Industry	413	580	734	1181	1481	1405	1486
Federal labs and other government	460	839	1173	803	747	693	623
Other jobs (including self-employed)	141	221	369	522	561	453	413

Research Involvement (in full-time S&E position)
Engaged primarily in research	1242	1907	2470	2406	3225	3087	3060
Supported by federal grants/contracts	1311	1871	2180	1956	n/a	2000	1858
Supported by HHS, NSF, and/or USDA	1150	1469	1827	1599	n/a	1778	1680

Table F.4F Career Progression of Life-Science PhDs from 26 High-Quality Institutions– US Citizens and Permanent Residents

	\multicolumn{7}{c}{Survey Year}
	1973	1977	1981	1985	1989	1993	1995
Panel A: Fraction of 1-2 Year Cohort							
Not in Full-time S&E Work Force							
Unemployed and seeking position	0.02	0.05	0.02	0.01	0.00	0.00	0.02
Part-time employed	0.03	0.03	0.03	0.03	0.01	0.02	0.02
Working outside science and engineering	0.01	0.02	0.01	0.05	0.01	0.03	0.01
Full-time employed in S&E field							
Tenure-track faculty position @ PhD Institutions	0.33	0.19	0.16	0.09	0.11	0.10	0.07
Tenure-track faculty position @ Other Inst	0.07	0.03	0.04	0.05	0.01	0.03	0.04
Postdoc total appointments (in any sector)	0.30	0.48	0.47	0.50	0.61	0.55	0.60
Other academic position	0.07	0.04	0.09	0.08	0.09	0.08	0.11
Industry	0.06	0.05	0.09	0.06	0.06	0.09	0.06
Federal labs and other government	0.06	0.06	0.04	0.07	0.05	0.07	0.03
Other jobs (including self-employed)	0.05	0.07	0.06	0.06	0.04	0.03	0.03
Research Involvement (in full-time S&E position)							
Engaged primarily in research	0.72	0.73	0.73	0.74	0.84	0.82	0.79
Supported by federal grants/contracts	0.67	0.68	0.61	0.51	n/a	0.49	0.55
Supported by HHS, NSF, and/or USDA	0.60	0.57	0.52	0.42	n/a	0.45	0.50
Panel B: Fraction of 3-4 Year Cohort							
Not in Full-time S&E Work Force							
Unemployed and seeking position	0.02	0.02	0.03	0.02	0.01	0.01	0.01
Part-time employed	0.03	0.02	0.02	0.03	0.04	0.04	0.03
Working outside science and engineering	0.03	0.02	0.01	0.01	0.02	0.03	0.05
Full-time employed in S&E field							
Tenure-track faculty position @ PhD Institutions	0.45	0.40	0.32	0.21	0.19	0.17	0.13
Tenure-track faculty position @ Other Inst	0.10	0.06	0.03	0.02	0.02	0.04	0.04
Postdoc total appointments (in any sector)	0.11	0.15	0.21	0.24	0.26	0.34	0.33
Other academic position	0.08	0.08	0.12	0.12	0.15	0.13	0.14
Industry	0.06	0.10	0.14	0.24	0.19	0.16	0.18
Federal labs and other government	0.07	0.09	0.07	0.04	0.06	0.05	0.07
Other jobs (including self-employed)	0.07	0.07	0.04	0.06	0.06	0.03	0.03
Research Involvement (in full-time S&E position)							
Engaged primarily in research	0.55	0.53	0.71	0.68	0.76	0.71	0.73
Supported by federal grants/contracts	0.65	0.61	0.54	0.48	n/a	0.48	0.50
Supported by HHS, NSF, and/or USDA	0.58	0.52	0.45	0.39	n/a	0.39	0.48

Appenidx F

	Survey Year							
	1973	1977	1981	1985	1989	1993	1995
Panel C: Fraction of 5-6 Year Cohort							
Not in Full-time S&E Work Force							
Unemployed and seeking position	0.01	0.01	0.00	0.00	0.01	0.00	0.01
Part-time employed	0.04	0.05	0.03	0.05	0.01	0.04	0.04
Working outside science and engineering	0.03	0.03	0.02	0.03	0.02	0.03	0.04
Full-time employed in S&E field							
Tenure-track faculty position @ PhD Institutions	0.51	0.49	0.44	0.36	0.31	0.31	0.24
Tenure-track faculty position @ Other Inst	0.11	0.11	0.05	0.03	0.06	0.03	0.05
Postdoc total appointments (in any sector)	0.01	0.06	0.06	0.10	0.14	0.13	0.17
Other academic position	0.04	0.05	0.12	0.08	0.12	0.09	0.18
Industry	0.09	0.09	0.09	0.20	0.17	0.19	0.18
Federal labs and other government	0.11	0.07	0.08	0.09	0.08	0.08	0.04
Other jobs (including self-employed)	0.04	0.05	0.10	0.06	0.07	0.11	0.04
Research Involvement (in full-time S&E position)							
Engaged primarily in research	0.48	0.58	0.61	0.56	0.72	0.77	0.72
Supported by federal grants/contracts	0.61	0.64	0.57	0.42	n/a	0.45	0.52
Supported by HHS, NSF, and/or USDA	0.52	0.53	0.52	0.38	n/a	0.40	0.46
Panel D: Fraction of 7-8 Year Cohort							
Not in Full-time S&E Work Force							
Unemployed and seeking position	0.01	0.04	0.01	0.01	0.00	0	0.02
Part-time employed	0.03	0.02	0.04	0.02	0.03	0.05	0.02
Working outside science and engineering	0.02	0.03	0.01	0.03	0.07	0.04	0.02
Full-time employed in S&E field							
Tenure-track faculty position @ PhD Institutions	0.54	0.47	0.52	0.45	0.36	0.35	0.40
Tenure-track faculty position @ Other Inst	0.11	0.12	0.04	0.03	0.02	0.05	0.02
Postdoc total appointments (in any sector)	0.02	0.04	0.02	0.02	0.04	0.06	0.06
Other academic position	0.03	0.03	0.05	0.09	0.12	0.11	0.09
Industry	0.07	0.08	0.11	0.17	0.24	0.20	0.19
Federal labs and other government	0.10	0.09	0.10	0.10	0.06	0.07	0.09
Other jobs (including self-employed)	0.07	0.08	0.10	0.09	0.06	0.07	0.10
Research Involvement (in full-time S&E position)							
Engaged primarily in research	0.45	0.48	0.49	0.64	0.70	0.66	0.72
Supported by federal grants/contracts	0.64	0.59	0.53	0.51	n/a	0.5	0.50
Supported by HHS, NSF, and/or USDA	0.60	0.49	0.41	0.49	n/a	0.49	0.47

	Survey Year						
	1973	1977	1981	1985	1989	1993	1995
Panel E: Fraction of 9-10 Year Cohort							
Not in Full-time S&E Work Force							
Unemployed and seeking position	0.02	0.01	0.02	0.01	0.01	0.01	0.01
Part-time employed	0.04	0.04	0.04	0.02	0.04	0.02	0.04
Working outside science and engineering	0.02	0.05	0.04	0.05	0.03	0.05	0.02
Full-time employed in S&E field							
Tenure-track faculty position @ PhD Institutions	0.48	0.51	0.54	0.41	0.44	0.37	0.45
Tenure-track faculty position @ Other Inst	0.14	0.11	0.06	0.08	0.04	0.07	0.05
Postdoc total appointments (in any sector)	0.00	0.01	0.01	0.02	0.01	0.03	0.01
Other academic position	0.02	0.02	0.04	0.10	0.07	0.09	0.08
Industry	0.09	0.10	0.11	0.12	0.22	0.16	0.21
Federal labs and other government	0.11	0.10	0.08	0.09	0.06	0.10	0.05
Other jobs (including self-employed)	0.07	0.05	0.06	0.11	0.08	0.09	0.08
Research Involvement (in full-time S&E position)							
Engaged primarily in research	0.42	0.48	0.58	0.51	0.64	0.61	0.68
Supported by federal grants/contracts	0.49	0.56	0.57	0.46	n/a	0.47	0.54
Supported by HHS, NSF, and/or USDA	0.47	0.48	0.54	0.41	n/a	0.41	0.53

Table F.4N Career Progression of Life-Science PhDs from 26 High-Quality Institutions– US Citizens and Permanent Residents

	Survey Year						
	1973	1977	1981	1985	1989	1993	1995
Panel A: Number in 1-2 Year Cohort							
Not in Full-time S&E Work Force							
Unemployed and seeking position	30	91	37	30	11		39
Part-time employed	54	57	83	62	28	44	44
Working outside science and engineering	21	43	17	125	28	68	20
Full-time employed in S&E field							
Tenure-track faculty position @ PhD Institutions	526	374	389	214	283	242	179
Tenure-track faculty position @ Other Inst	112	58	87	125	24	58	100
Postdoc total appointments (in any sector)	476	963	1166	1194	1531	1326	1555
Other academic position	116	81	211	182	217	193	296
Industry	91	98	227	156	145	211	166
Federal labs and other government	101	118	99	179	127	169	88
Other jobs (including self-employed)	75	131	150	134	101	83	90
Research Involvement (in full-time S&E position)							
Engaged primarily in research	1079	1337	1710	1607	2033	1874	1943
Supported by federal grants/contracts	1007	1235	1413	1108	n/a	1128	1369
Supported by HHS, NSF, and/or USDA	899	1037	1205	910	n/a	1033	1239
Panel B: Number in 3-4 Year Cohort							
Not in Full-time S&E Work Force							
Unemployed and seeking position	33	31	80	43	28	23	27
Part-time employed	41	31	52	73	102	98	75
Working outside science and engineering	46	35	12	31	46	79	110
Full-time employed in S&E field							
Tenure-track faculty position @ PhD Institutions	706	745	729	536	454	378	312
Tenure-track faculty position @ Other Inst	153	111	77	57	49	90	85
Postdoc total appointments (in any sector)	168	283	495	607	622	773	793
Other academic position	121	141	274	290	350	302	327
Industry	91	190	334	602	452	360	437
Federal labs and other government	109	160	157	97	141	117	176
Other jobs (including self-employed)	103	122	102	160	131	57	79
Research Involvement (in full-time S&E position)							
Engaged primarily in research	796	929	1544	1588	1674	1470	1619
Supported by federal grants/contracts	937	1062	1165	1124	n/a	992	1112
Supported by HHS, NSF, and/or USDA	841	907	968	915	n/a	816	1062

	Survey Year						
	1973	1977	1981	1985	1989	1993	1995
Panel C: Number in 5-6 Year Cohort							
Not in Full-time S&E Work Force							
Unemployed and seeking position	19	12	9	8	20	8	25
Part-time employed	64	95	71	116	32	86	89
Working outside science and engineering	42	55	35	72	49	72	92
Full-time employed in S&E field							
Tenure-track faculty position @ PhD Institutions	733	904	946	926	719	713	513
Tenure-track faculty position @ Other Inst	165	197	115	80	150	61	102
Postdoc total appointments (in any sector)	21	104	135	254	328	297	381
Other academic position	53	92	261	201	292	200	396
Industry	123	159	196	514	402	434	400
Federal labs and other government	155	121	162	230	179	184	98
Other jobs (including self-employed)	64	97	225	160	170	252	84
Research Involvement (in full-time S&E position)							
Engaged primarily in research	627	964	1239	1322	1607	1657	1431
Supported by federal grants/contracts	801	1066	1170	993	n/a	953	1019
Supported by HHS, NSF, and/or USDA	678	893	1054	888	n/a	862	905
Panel D: Number in 7-8 Year Cohort							
Not in Full-time S&E Work Force							
Unemployed and seeking position	11	62	20	21			44
Part-time employed	31	34	78	36	71	120	36
Working outside science and engineering	24	50	27	58	189	91	49
Full-time employed in S&E field							
Tenure-track faculty position @ PhD Institutions	582	766	1078	976	925	830	900
Tenure-track faculty position @ Other Inst	120	200	83	71	44	126	38
Postdoc total appointments (in any sector)	21	57	34	45	112	152	143
Other academic position	28	45	110	186	321	253	208
Industry	74	129	233	360	623	483	429
Federal labs and other government	113	144	198	214	147	155	198
Other jobs (including self-employed)	75	132	199	188	147	162	231
Research Involvement (in full-time S&E position)							
Engaged primarily in research	455	709	939	1309	1613	1424	1545
Supported by federal grants/contracts	648	870	1025	1047	n/a	1088	1081
Supported by HHS, NSF, and/or USDA	604	727	797	1001	n/a	1050	1017

	\multicolumn{7}{c}{Survey Year}
	1973	1977	1981	1985	1989	1993	1995
Panel E: Number in 9-10 Year Cohort							
Not in Full-time S&E Work Force							
Unemployed and seeking position	19	11	33	12	26	34	18
Part-time employed	38	59	80	55	120	55	103
Working outside science and engineering	25	68	67	101	77	123	48
Full-time employed in S&E field							
Tenure-track faculty position @ PhD Institutions	502	753	988	913	1218	884	1069
Tenure-track faculty position @ Other Inst	149	157	119	175	104	175	117
Postdoc total appointments (in any sector)	.		15	10	45	37	63	24
Other academic position	17	31	70	224	202	205	195
Industry	96	153	208	261	604	391	508
Federal labs and other government	118	155	146	205	163	248	127
Other jobs (including self-employed)	78	81	111	236	223	204	183
Research Involvement (in full-time S&E position)							
Engaged primarily in research	405	641	953	1054	1639	1325	1507
Supported by federal grants/contracts	474	749	938	942	n/a	1014	1197
Supported by HHS, NSF, and/or USDA	447	644	899	845	n/a	895	1173

Table F.5F   Career Progression of Life-Science PhDs from Other Institutions–US Citizens and Permanent Residents

	Survey Year						
	1973	1977	1981	1985	1989	1993	1995
Panel A: Fraction of 1-2 Year Cohort							
Not in Full-time S&E Work Force							
Unemployed and seeking position	0.02	0.01	0.02	0.03	0.02	0.01	0.02
Part-time employed	0.02	0.01	0.02	0.03	0.04	0.03	0.02
Working outside science and engineering	0.02	0.03	0.01	0.02	0.02	0.03	0.04
Full-time employed in S&E field							
Tenure-track faculty position @ PhD Institutions	0.23	0.20	0.17	0.13	0.12	0.10	0.10
Tenure-track faculty position @ Other Inst	0.17	0.11	0.08	0.06	0.05	0.07	0.05
Postdoc total appointments (in any sector)	0.19	0.34	0.41	0.39	0.43	0.41	0.50
Other academic position	0.08	0.04	0.06	0.10	0.06	0.12	0.11
Industry	0.09	0.12	0.12	0.13	0.14	0.13	0.08
Federal labs and other government	0.12	0.10	0.06	0.05	0.08	0.06	0.04
Other jobs (including self-employed)	0.05	0.03	0.04	0.05	0.05	0.03	0.03
Research Involvement (in full-time S&E position)							
Engaged primarily in research	0.50	0.54	0.67	0.65	0.73	0.66	0.67
Supported by federal grants/contracts	0.48	0.47	0.55	0.43	n/a	0.43	0.47
Supported by HHS, NSF, and/or USDA	0.39	0.37	0.47	0.33	n/a	0.36	0.43
Panel B: Fraction of 3-4 Year Cohort							
Not in Full-time S&E Work Force							
Unemployed and seeking position	0.01	0.03	0.01	0.01	0.01	0.01	0.02
Part-time employed	0.02	0.01	0.01	0.02	0.03	0.05	0.03
Working outside science and engineering	0.03	0.03	0.03	0.01	0.02	0.05	0.05
Full-time employed in S&E field							
Tenure-track faculty position @ PhD Institutions	0.38	0.28	0.28	0.25	0.21	0.17	0.13
Tenure-track faculty position @ Other Inst	0.17	0.15	0.10	0.07	0.03	0.07	0.09
Postdoc total appointments (in any sector)	0.04	0.10	0.17	0.19	0.21	0.26	0.28
Other academic position	0.05	0.05	0.07	0.08	0.11	0.09	0.12
Industry	0.12	0.15	0.17	0.20	0.20	0.18	0.16
Federal labs and other government	0.13	0.11	0.10	0.09	0.11	0.07	0.08
Other jobs (including self-employed)	0.06	0.09	0.05	0.06	0.06	0.05	0.04
Research Involvement (in full-time S&E position)							
Engaged primarily in research	0.41	0.42	0.61	0.58	0.67	0.66	0.59
Supported by federal grants/contracts	0.47	0.42	0.41	0.39	n/a	0.38	0.43
Supported by HHS, NSF, and/or USDA	0.38	0.34	0.34	0.32	n/a	0.34	0.37

	Survey Year						
	1973	1977	1981	1985	1989	1993	1995
Panel C: Fraction of 5-6 Year Cohort							
Not in Full-time S&E Work Force							
Unemployed and seeking position	0.01	0.02	0.03	0.01	0.01	0.02	0.02
Part-time employed	0.03	0.02	0.01	0.03	0.01	0.04	0.05
Working outside science and engineering	0.02	0.04	0.02	0.04	0.03	0.05	0.06
Full-time employed in S&E field							
Tenure-track faculty position @ PhD Institutions	0.39	0.34	0.30	0.30	0.26	0.25	0.22
Tenure-track faculty position @ Other Inst	0.16	0.16	0.12	0.09	0.08	0.07	0.09
Postdoc total appointments (in any sector)	0.02	0.04	0.07	0.10	0.09	0.12	0.13
Other academic position	0.03	0.03	0.05	0.06	0.10	0.07	0.10
Industry	0.11	0.12	0.19	0.20	0.26	0.24	0.19
Federal labs and other government	0.17	0.15	0.15	0.10	0.09	0.10	0.09
Other jobs (including self-employed)	0.06	0.08	0.06	0.06	0.07	0.04	0.05
Research Involvement (in full-time S&E position)							
Engaged primarily in research	0.43	0.38	0.47	0.52	0.60	0.58	0.62
Supported by federal grants/contracts	0.41	0.42	0.37	0.33	n/a	0.34	0.35
Supported by HHS, NSF, and/or USDA	0.35	0.33	0.31	0.27	n/a	0.30	0.31
Panel D: Fraction of 7-8 Year Cohort							
Not in Full-time S&E Work Force							
Unemployed and seeking position	0.01	0.01	0.01	0.01	0.01	0.01	0.01
Part-time employed	0.02	0.02	0.02	0.02	0.04	0.03	0.03
Working outside science and engineering	0.03	0.03	0.03	0.05	0.04	0.06	0.06
Full-time employed in S&E field							
Tenure-track faculty position @ PhD Institutions	0.45	0.43	0.35	0.35	0.31	0.29	0.29
Tenure-track faculty position @ Other Inst	0.16	0.14	0.14	0.10	0.07	0.06	0.09
Postdoc total appointments (in any sector)	0.01	0.03	0.03	0.02	0.05	0.05	0.06
Other academic position	0.02	0.03	0.04	0.05	0.08	0.10	0.07
Industry	0.13	0,12	0.19	0.22	0.24	0.22	0.23
Federal labs and other government	0.14	0.14	0.11	0.11	0.09	0.12	0.10
Other jobs (including self-employed)	0.04	0.07	0.09	0.07	0.08	0.06	0.06
Research Involvement (in full-time S&E position)							
Engaged primarily in research	0.41	0.39	0.40	0.46	0.58	0.59	0.56
Supported by federal grants/contracts	0.46	0.41	0.33	0.32	n/a	0.34	0.34
Supported by HHS, NSF, and/or USDA	0.37	0.34	0.27	0.26	n/a	0.29	0.30

	\multicolumn{7}{c}{Survey Year}
	1973	1977	1981	1985	1989	1993	1995
Panel E: Fraction of 9-10 Year Cohort							
Not in Full-time S&E Work Force							
Unemployed and seeking position	0.00	0.02	0.02	0.01	0.00	0.01	0.00
Part-time employed	0.01	0.02	0.02	0.03	0.03	0.02	0.02
Working outside science and engineering	0.04	0.04	0.04	0.04	0.04	0.07	0.05
Full-time employed in S&E field							
Tenure-track faculty position @ PhD Institutions	0.47	0.39	0.34	0.32	0.33	0.28	0.29
Tenure-track faculty position @ Other Inst	0.14	0.16	0.17	0.11	0.07	0.09	0.05
Postdoc total appointments (in any sector)	0.00	0.01	0.01	0.02	0.02	0.03	0.04
Other academic position	0.01	0.01	0.05	0.04	0.05	0.06	0.10
Industry	0.13	0.12	0.11	0.20	0.23	0.24	0.24
Federal labs and other government	0.15	0.18	0.19	0.14	0.13	0.12	0.13
Other jobs (including self-employed)	0.03	0.05	0.06	0.08	0.09	0.06	0.07
Research Involvement (in full-time S&E position)							
Engaged primarily in research	0.41	0.42	0.35	0.41	0.48	0.56	0.54
Supported by federal grants/contracts	0.41	0.39	0.30	0.33	n/a	0.33	0.32
Supported by HHS, NSF, and/or USDA	0.35	0.29	0.23	0.26	n/a	0.30	0.28

Appenidx F

Table F.5N Career Progression of Life-Science PhDs from Other Institutions–US Citizens and Permanent Residents

	\multicolumn{7}{c}{Survey Year}
	1973	1977	1981	1985	1989	1993	1995
Panel A: Number in 1-2 Year Cohort							
Not in Full-time S&E Work Force							
Unemployed and seeking position	114	84	139	190	100	95	172
Part-time employed	141	71	139	174	272	177	158
Working outside science and engineering	115	177	61	137	138	223	299
Full-time employed in S&E field							
Tenure-track faculty position @ PhD Institutions	1410	1099	958	742	719	633	674
Tenure-track faculty position @ Other Inst	1046	625	443	385	291	428	369
Postdoc total appointments (in any sector)	1183	1884	2281	2304	2610	2626	3425
Other academic position	496	241	328	589	357	775	735
Industry	568	685	679	799	868	864	579
Federal labs and other government	763	565	331	305	470	382	271
Other jobs (including self-employed)	337	190	208	304	280	205	203
Research Involvement (in full-time S&E position)							
Engaged primarily in research	2884	2877	3487	3525	4060	3925	4188
Supported by federal grants/contracts	2800	2484	2900	2325	n/a	2517	2939
Supported by HHS, NSF, and/or USDA	2289	1976	2451	1817	n/a	2146	2660
Panel B: Number in 3-4 Year Cohort							
Not in Full-time S&E Work Force							
Unemployed and seeking position	37	179	47	89	47	41	146
Part-time employed	111	82	71	132	193	296	213
Working outside science and engineering	133	150	145	63	114	308	348
Full-time employed in S&E field							
Tenure-track faculty position @ PhD Institutions	1978	1591	1572	1481	1240	1014	850
Tenure-track faculty position @ Other Inst	856	872	554	434	198	452	572
Postdoc total appointments (in any sector)	216	573	961	1135	1208	1579	1771
Other academic position	258	277	369	483	624	569	779
Industry	615	817	955	1202	1141	1059	995
Federal labs and other government	649	605	579	557	661	435	480
Other jobs (including self-employed)	287	483	281	361	356	298	234
Research Involvement (in full-time S&E position)							
Engaged primarily in research	1980	2195	3214	3273	3645	3589	3376
Supported by federal grants/contracts	2286	2166	2139	2186	n/a	2054	2439
Supported by HHS, NSF, and/or USDA	1839	1749	1775	1804	n/a	1858	2079

	Survey Year						
	1973	1977	1981	1985	1989	1993	1995
Panel C: Number in 5-6 Year Cohort							
Not in Full-time S&E Work Force							
Unemployed and seeking position	24	120	143	46	72	97	141
Part-time employed	102	122	78	185	85	222	309
Working outside science and engineering	60	241	137	217	197	269	368
Full-time employed in S&E field							
Tenure-track faculty position @ PhD Institutions	1478	2184	1645	1647	1526	1454	1303
Tenure-track faculty position @ Other Inst	611	1007	668	525	501	415	535
Postdoc total appointments (in any sector)	73	279	381	561	511	704	790
Other academic position	124	215	277	327	569	421	582
Industry	426	746	1018	1143	1572	1386	1172
Federal labs and other government	650	944	800	570	555	594	535
Other jobs (including self-employed)	236	516	354	357	393	253	312
Research Involvement (in full-time S&E position)							
Engaged primarily in research	1559	2236	2437	2693	3384	3045	3252
Supported by federal grants/contracts	1470	2448	1897	1695	n/a	1783	1831
Supported by HHS, NSF, and/or USDA	1248	1921	1571	1383	n/a	1592	1607
Panel D: Number in 7-8 Year Cohort							
Not in Full-time S&E Work Force							
Unemployed and seeking position	25	33	49	73	32	66	69
Part-time employed	44	85	91	130	229	188	191
Working outside science and engineering	97	149	140	263	240	372	316
Full-time employed in S&E field							
Tenure-track faculty position @ PhD Institutions	1314	2230	1821	1949	1938	1663	1621
Tenure-track faculty position @ Other Inst	461	730	715	560	444	319	530
Postdoc total appointments (in any sector)	24	134	151	111	325	308	322
Other academic position	49	138	223	305	508	562	386
Industry	377	643	1025	1223	1502	1254	1305
Federal labs and other government	401	726	597	615	583	715	572
Other jobs (including self-employed)	126	339	450	405	506	342	326
Research Involvement (in full-time S&E position)							
Engaged primarily in research	1116	1931	2010	2388	3345	3067	2856
Supported by federal grants/contracts	1268	2038	1645	1669	n/a	1774	1745
Supported by HHS, NSF, and/or USDA	1011	1674	1348	1337	n/a	1521	1540

Appenidx F

	Survey Year						
	1973	1977	1981	1985	1989	1993	1995
Panel E: Number in 9-10 Year Cohort							
Not in Full-time S&E Work Force							
Unemployed and seeking position	11	66	95	69	15	77	26
Part-time employed	34	70	100	177	157	129	139
Working outside science and engineering	90	156	219	196	219	436	292
Full-time employed in S&E field							
Tenure-track faculty position @ PhD Institutions	1170	1514	2068	1652	1787	1704	1656
Tenure-track faculty position @ Other Inst	352	634	1037	568	404	554	305
Postdoc total appointments (in any sector)	7	49	49	87	134	201	204
Other academic position	30	32	327	201	289	369	550
Industry	334	464	639	1060	1280	1470	1374
Federal labs and other government	367	722	1114	734	732	716	755
Other jobs (including self-employed)	84	196	361	440	468	360	405
Research Involvement (in full-time S&E position)							
Engaged primarily in research	952	1503	1950	1950	2453	3009	2838
Supported by federal grants/contracts	972	1393	1663	1542	n/a	1798	1671
Supported by HHS, NSF, and/or USDA	819	1055	1308	1237	n/a	1601	1491

Table F.6F Career Progression of Nonbiomedical Life-Science PhDs–US Citizens and Permanent Residents

	Survey Year						
	1973	1977	1981	1985	1989	1993	1995
Panel A: Fraction of 1-2 Year Cohort							
Not in Full-time S&E Work Force							
Unemployed and seeking position	0.01	0.02	0.03	0.05	0.01	0.02	0.01
Part-time employed	0.02	0.01	0.02	0.05	0.07	0.02	0.04
Working outside science and engineering	0.02	0.03	0.01	0.03	0.02	0.04	0.08
Full-time employed in S&E field							
Tenure-track faculty position @ PhD Institutions	0.27	0.27	0.30	0.24	0.20	0.12	0.16
Tenure-track faculty position @ Other Inst	0.24	0.11	0.13	0.07	0.06	0.11	0.09
Postdoc total appointments (in any sector)	0.06	0.14	0.14	0.19	0.22	0.30	0.33
Other academic position	0.06	0.05	0.04	0.12	0.03	0.11	0.07
Industry	0.11	0.18	0.18	0.14	0.23	0.13	0.11
Federal labs and other government	0.19	0.18	0.11	0.10	0.12	0.12	0.05
Other jobs (including self-employed)	0.02	0.03	0.03	0.02	0.04	0.03	0.06
Research Involvement (in full-time S&E position)							
Engaged primarily in research	0.41	0.45	0.51	0.61	0.69	0.62	0.62
Supported by federal grants/contracts	0.39	0.38	0.51	0.37	n/a	0.38	0.41
Supported by HHS, NSF, and/or USDA	0.29	0.26	0.31	0.25	n/a	0.28	0.31
Panel B: Fraction of 3-4 Year Cohort							
Not in Full-time S&E Work Force							
Unemployed and seeking position	0.00	0.04	0.01	0.02	0.01	0.01	0.02
Part-time employed	0.01	0.01	0.01	0.05	0.04	0.07	0.03
Working outside science and engineering	0.01	0.02	0.04	0.01	0.02	0.04	0.04
Full-time employed in S&E field							
Tenure-track faculty position @ PhD Institutions	0.43	0.29	0.24	0.37	0.29	0.22	0.13
Tenure-track faculty position @ Other Inst	0.17	0.16	0.13	0.03	0.04	0.06	0.11
Postdoc total appointments (in any sector)	0.02	0.02	0.04	0.09	0.07	0.15	0.22
Other academic position	0.03	0.05	0.09	0.09	0.08	0.10	0.13
Industry	0.15	0.21	0.26	0.22	0.20	0.19	0.15
Federal labs and other government	0.17	0.18	0.16	0.10	0.19	0.13	0.13
Other jobs (including self-employed)	0.01	0.02	0.03	0.02	0.05	0.04	0.05
Research Involvement (in full-time S&E position)							
Engaged primarily in research	0.38	0.33	0.53	0.57	0.58	0.59	0.52
Supported by federal grants/contracts	0.43	0.36	0.33	0.45	n/a	0.22	0.44
Supported by HHS, NSF, and/or USDA	0.35	0.31	0.20	0.35	n/a	0.15	0.34

	Survey Year						
	1973	1977	1981	1985	1989	1993	1995

Panel C: Fraction of 5-6 Year Cohort

Not in Full-time S&E Work Force

Unemployed and seeking position	0.01	0.01	0.02	0.00	0.01	0.01	0.05
Part-time employed	0.03	0.02	0.01	0.02	0.01	0.06	0.07
Working outside science and engineering	0.02	0.04	0.01	0.06	0.01	0.06	0.04

Full-time employed in S&E field

Tenure-track faculty position @ PhD Institutions	0.34	0.30	0.27	0.33	0.37	0.29	0.24
Tenure-track faculty position @ Other Inst	0.20	0.21	0.14	0.14	0.09	0.07	0.03
Postdoc total appointments (in any sector)	0.01	0.03	0.00	0.05	0.03	0.05	0.12
Other academic position	0.01	0.01	0.07	0.04	0.07	0.05	0.1
Industry	0.10	0.12	0.18	0.18	0.21	0.24	0.19
Federal labs and other government	0.27	0.23	0.26	0.15	0.16	0.11	0.15
Other jobs (including self-employed)	0.02	0.04	0.04	0.05	0.04	0.06	0.02

Research Involvement (in full-time S&E position)

Engaged primarily in research	0.45	0.31	0.45	0.43	0.63	0.52	0.69
Supported by federal grants/contracts	0.37	0.41	0.28	0.33	n/a	0.33	0.32
Supported by HHS, NSF, and/or USDA	0.31	0.31	0.22	0.23	n/a	0.25	0.23

Panel D: Fraction of 7-8 Year Cohort

Not in Full-time S&E Work Force

Unemployed and seeking position	0.00	0.01	0.01	0.02	0.00	0.01	0.01
Part-time employed	0.00	0.01	0.00	0.01	0.05	0.02	0.02
Working outside science and engineering	0.04	0.02	0.02	0.04	0.05	0.09	0.09

Full-time employed in S&E field

Tenure-track faculty position @ PhD Institutions	0.49	0.45	0.34	0.31	0.33	0.26	0.27
Tenure-track faculty position @ Other Inst	0.15	0.13	0.15	0.11	0.05	0.08	0.11
Postdoc total appointments (in any sector)	0.00	0.01	0.00	0.02	0.02	0.03	0.03
Other academic position	0.01	0.02	0.03	0.03	0.09	0.08	0.05
Industry	0.11	0.13	0.23	0.25	0.25	0.18	0.22
Federal labs and other government	0.18	0.18	0.18	0.17	0.13	0.20	0.12
Other jobs (including self-employed)	0.02	0.03	0.04	0.06	0.03	0.05	0.06

Research Involvement (in full-time S&E position)

Engaged primarily in research	0.44	0.38	0.31	0.45	0.53	0.6	0.52
Supported by federal grants/contracts	0.52	0.42	0.30	0.29	n/a	0.29	0.37
Supported by HHS, NSF, and/or USDA	0.43	0.34	0.23	0.19	n/a	0.21	0.32

	Survey Year						
	1973	1977	1981	1985	1989	1993	1995
Panel E: Fraction of 9-10 Year Cohort							
Not in Full-time S&E Work Force							
Unemployed and seeking position	0.01	0.01	0.01	0.01	0.00	0.02	0.00
Part-time employed	0.01	0.01	0.01	0.03	0.02	0.02	0.02
Working outside science and engineering	0.03	0.04	0.05	0.05	0.03	0.07	0.05
Full-time employed in S&E field							
Tenure-track faculty position @ PhD Institutions	0.47	0.31	0.26	0.26	0.37	0.30	0.28
Tenure-track faculty position @ Other Inst	0.16	0.19	0.26	0.13	0.10	0.12	0.06
Postdoc total appointments (in any sector)	0.00	0.00	0.00	0.01	0.04	0.02	0.03
Other academic position	0.01	0.02	0.04	0.05	0.05	0.05	0.10
Industry	0.10	0.09	0.08	0.18	0.14	0.19	0.18
Federal labs and other government	0.19	0.30	0.25	0.25	0.17	0.18	0.20
Other jobs (including self-employed)	0.01	0.02	0.03	0.04	0.08	0.04	0.07
Research Involvement (in full-time S&E position)							
Engaged primarily in research	0.45	0.43	0.32	0.41	0.46	0.51	0.57
Supported by federal grants/contracts	0.39	0.36	0.27	0.28	n/a	0.29	0.32
Supported by HHS, NSF, and/or USDA	0.33	0.24	0.20	0.23	n/a	0.21	0.28

Table F.6N   Career Progression of Nonbiomedical Life-Science PhDs–US Citizens and Permanent Residents

	Survey Year						
	1973	1977	1981	1985	1989	1993	1995
Panel A: Number in 1-2 Year Cohort							
Not in Full-time S&E Work Force							
Unemployed and seeking position	23	31	66	105	31	43	12
Part-time employed	38	11	48	105	146	39	78
Working outside science and engineering	45	56	26	79	37	76	151
Full-time employed in S&E field							
Tenure-track faculty position @ PhD Institutions	615	547	573	562	426	259	307
Tenure-track faculty position @ Other Inst	542	218	243	163	135	237	178
Postdoc total appointments (in any sector)	138	290	280	438	460	643	632
Other academic position	125	107	78	270	57	228	131
Industry	258	364	344	313	492	278	221
Federal labs and other government	423	374	222	227	263	256	103
Other jobs (including self-employed)	49	57	55	49	89	60	111
Research Involvement (in full-time S&E position)							
Engaged primarily in research	890	879	914	1238	1324	1211	1047
Supported by federal grants/contracts	833	751	908	758	n/a	743	688
Supported by HHS, NSF, and/or USDA	615	512	561	509	n/a	542	520
Panel B: Number in 3-4 Year Cohort							
Not in Full-time S&E Work Force							
Unemployed and seeking position	4	74	25	34	13	17	32
Part-time employed	17	23	21	102	84	129	59
Working outside science and engineering	22	41	75	33	42	69	88
Full-time employed in S&E field							
Tenure-track faculty position @ PhD Institutions	849	609	508	818	635	426	270
Tenure-track faculty position @ Other Inst	338	328	264	71	92	108	230
Postdoc total appointments (in any sector)	35	41	86	204	159	291	462
Other academic position	59	108	180	205	171	185	267
Industry	298	442	533	493	434	355	305
Federal labs and other government	337	366	341	229	419	254	271
Other jobs (including self-employed)	17	35	56	38	112	81	97
Research Involvement (in full-time S&E position)							
Engaged primarily in research	737	638	1051	1167	1163	1007	997
Supported by federal grants/contracts	823	704	657	918	n/a	382	843
Supported by HHS, NSF, and/or USDA	671	601	387	720	n/a	254	639

	Survey Year						
	1973	1977	1981	1985	1989	1993	1995

Panel C: Number in 5-6 Year Cohort

Not in Full-time S&E Work Force

Unemployed and seeking position	12	21	35	0	25	18	95
Part-time employed	45	36	25	34	23	124	124
Working outside science and engineering	30	94	13	105	26	117	70

Full-time employed in S&E field

Tenure-track faculty position @ PhD Institutions	519	688	527	620	763	568	447
Tenure-track faculty position @ Other Inst	312	487	268	267	180	142	48
Postdoc total appointments (in any sector)	16	59	2	102	57	108	213
Other academic position	14	26	131	70	141	94	179
Industry	154	274	341	339	436	468	348
Federal labs and other government	416	514	493	283	337	216	279
Other jobs (including self-employed)	28	81	86	86	93	110	38

Research Involvement (in full-time S&E position)

Engaged primarily in research	663	662	834	758	1256	883	1064
Supported by federal grants/contracts	539	882	517	580	n/a	570	493
Supported by HHS, NSF, and/or USDA	458	650	399	409	n/a	429	354

Panel D: Number in 7-8 Year Cohort

Not in Full-time S&E Work Force

Unemployed and seeking position	0	24	18	31	4	30	25
Part-time employed	5	24	5	23	124	51	35
Working outside science and engineering	55	37	33	77	121	178	161

Full-time employed in S&E field

Tenure-track faculty position @ PhD Institutions	629	922	617	636	782	549	489
Tenure-track faculty position @ Other Inst	190	264	268	217	117	168	206
Postdoc total appointments (in any sector)	5	29	2	37	44	54	58
Other academic position	8	47	55	66	203	169	96
Industry	138	265	429	507	595	369	387
Federal labs and other government	233	369	330	356	295	421	219
Other jobs (including self-employed)	32	60	70	116	59	99	116

Research Involvement (in full-time S&E position) 2088

Engaged primarily in research	545	746	557	874	1120	1092	823
Supported by federal grants/contracts	637	820	524	563	n/a	525	577
Supported by HHS, NSF, and/or USDA	528	656	405	374	n/a	380	510

	\multicolumn{7}{c}{Survey Year}
	1973	1977	1981	1985	1989	1993	1995
Panel E: Number in 9-10 Year Cohort							
Not in Full-time S&E Work Force							
Unemployed and seeking position	13	21	24	12	0	43	0
Part-time employed	11	12	31	64	46	37	41
Working outside science and engineering	43	56	124	91	58	159	99
Full-time employed in S&E field							
Tenure-track faculty position @ PhD Institutions	587	456	674	483	733	671	547
Tenure-track faculty position @ Other Inst	205	280	658	240	197	276	117
Postdoc total appointments (in any sector)	3	5	6	10	74	34	53
Other academic position	18	23	101	86	90	113	198
Industry	128	137	216	340	274	428	349
Federal labs and other government	236	434	650	458	330	397	393
Other jobs (including self-employed)	18	34	75	74	164	83	138
Research Involvement (in full-time S&E position)							
Engaged primarily in research	533	589	751	693	865	1013	1024
Supported by federal grants/contracts	465	498	645	472	n/a	575	571
Supported by HHS, NSF, and/or USDA	393	330	473	394	n/a	430	507

Table F.7F Career Progression of Biomedical Life-Science PhDs–US Citizens and Permanent Residents

	\multicolumn{7}{c}{Survey Year}
	1973	1977	1981	1985	1989	1993	1995
Panel A: Fraction of 1-2 Year Cohort							
Not in Full-time S&E Work Force							
Unemployed and seeking position	0.02	0.03	0.02	0.02	0.01	0.01	0.03
Part-time employed	0.03	0.02	0.03	0.02	0.02	0.03	0.02
Working outside science and engineering	0.02	0.03	0.01	0.03	0.02	0.03	0.02
Full-time employed in S&E field							
Tenure-track faculty position @ PhD Institutions	0.24	0.17	0.13	0.07	0.09	0.09	0.07
Tenure-track faculty position @ Other Inst	0.11	0.08	0.05	0.06	0.03	0.04	0.04
Postdoc total appointments (in any sector)	0.28	0.46	0.52	0.51	0.57	0.49	0.58
Other academic position	0.09	0.04	0.08	0.08	0.08	0.11	0.12
Industry	0.07	0.08	0.09	0.11	0.08	0.12	0.07
Federal labs and other government	0.08	0.06	0.03	0.04	0.05	0.04	0.03
Other jobs (including self-employed)	0.07	0.05	0.05	0.06	0.05	0.03	0.02
Research Involvement (in full-time S&E position)							
Engaged primarily in research	0.60	0.65	0.74	0.70	0.78	0.74	0.72
Supported by federal grants/contracts	0.58	0.58	0.59	0.48	n/a	0.47	0.51
Supported by HHS, NSF, and/or USDA	0.50	0.49	0.54	0.40	n/a	0.42	0.48
Panel B: Fraction of 3-4 Year Cohort							
Not in Full-time S&E Work Force							
Unemployed and seeking position	0.01	0.03	0.02	0.02	0.01	0.01	0.02
Part-time employed	0.03	0.02	0.02	0.02	0.04	0.04	0.03
Working outside science and engineering	0.03	0.03	0.01	0.01	0.02	0.05	0.05
Full-time employed in S&E field							
Tenure-track faculty position @ PhD Institutions	0.39	0.32	0.31	0.19	0.18	0.15	0.13
Tenure-track faculty position @ Other Inst	0.14	0.12	0.06	0.07	0.03	0.07	0.06
Postdoc total appointments (in any sector)	0.07	0.15	0.24	0.25	0.28	0.32	0.31
Other academic position	0.07	0.06	0.08	0.09	0.13	0.11	0.12
Industry	0.09	0.10	0.13	0.21	0.19	0.17	0.17
Federal labs and other government	0.09	0.07	0.07	0.07	0.06	0.05	0.06
Other jobs (including self-employed)	0.08	0.11	0.06	0.08	0.06	0.04	0.03
Research Involvement (in full-time S&E position)							
Engaged primarily in research	0.47	0.49	0.68	0.62	0.74	0.70	0.67
Supported by federal grants/contracts	0.55	0.50	0.48	0.40	n/a	0.46	0.45
Supported by HHS, NSF, and/or USDA	0.46	0.41	0.43	0.34	n/a	0.42	0.42

Appenidx F

	Survey Year						
	1973	1977	1981	1985	1989	1993	1995
Panel C: Fraction of 5-6 Year Cohort							
Not in Full-time S&E Work Force							
Unemployed and seeking position	0.01	0.02	0.02	0.01	0.01	0.01	0.01
Part-time employed	0.03	0.03	0.02	0.04	0.02	0.03	0.04
Working outside science and engineering	0.02	0.03	0.03	0.03	0.04	0.04	0.06
Full-time employed in S&E field							
Tenure-track faculty position @ PhD Institutions	0.46	0.40	0.36	0.31	0.24	0.26	0.21
Tenure-track faculty position @ Other Inst	0.13	0.12	0.09	0.05	0.08	0.05	0.09
Postdoc total appointments (in any sector)	0.02	0.05	0.09	0.11	0.13	0.15	0.15
Other academic position	0.04	0.05	0.07	0.07	0.12	0.09	0.13
Industry	0.11	0.11	0.15	0.21	0.25	0.22	0.19
Federal labs and other government	0.11	0.09	0.08	0.08	0.06	0.09	0.06
Other jobs (including self-employed)	0.07	0.09	0.09	0.07	0.08	0.06	0.06
Research Involvement (in full-time S&E position)							
Engaged primarily in research	0.44	0.47	0.54	0.57	0.64	0.67	0.64
Supported by federal grants/contracts	0.50	0.48	0.48	0.37	n/a	0.38	0.42
Supported by HHS, NSF, and/or USDA	0.43	0.40	0.42	0.33	n/a	0.36	0.38
Panel D: Fraction of 7-8 Year Cohort							
Not in Full-time S&E Work Force							
Unemployed and seeking position	0.01	0.01	0.01	0.01	0.00	0.01	0.01
Part-time employed	0.03	0.02	0.03	0.02	0.03	0.04	0.03
Working outside science and engineering	0.02	0.03	0.02	0.04	0.05	0.05	0.03
Full-time employed in S&E field							
Tenure-track faculty position @ PhD Institutions	0.47	0.43	0.42	0.40	0.32	0.32	0.33
Tenure-track faculty position @ Other Inst	0.14	0.14	0.10	0.07	0.06	0.05	0.06
Postdoc total appointments (in any sector)	0.01	0.03	0.03	0.02	0.06	0.07	0.07
Other academic position	0.03	0.03	0.05	0.07	0.10	0.11	0.08
Industry	0.12	0.11	0.15	0.19	0.23	0.23	0.22
Federal labs and other government	0.10	0.10	0.08	0.08	0.07	0.07	0.09
Other jobs (including self-employed)	0.06	0.09	0.11	0.08	0.09	0.07	0.07
Research Involvement (in full-time S&E position)							
Engaged primarily in research	0.41	0.42	0.46	0.54	0.64	0.62	0.63
Supported by federal grants/contracts	0.51	0.47	0.42	0.41	n/a	0.43	0.40
Supported by HHS, NSF, and/or USDA	0.43	0.39	0.34	0.37	n/a	0.40	0.36

	Survey Year						
	1973	1977	1981	1985	1989	1993	1995
Panel E: Fraction of 9-10 Year Cohort							
Not in Full-time S&E Work Force							
Unemployed and seeking position	0.01	0.01	0.02	0.01	0.01	0.01	0.01
Part-time employed	0.03	0.03	0.03	0.03	0.04	0.02	0.03
Working outside science and engineering	0.03	0.04	0.03	0.04	0.04	0.06	0.04
Full-time employed in S&E field							
Tenure-track faculty position @ PhD Institutions	0.48	0.46	0.45	0.37	0.36	0.31	0.35
Tenure-track faculty position @ Other Inst	0.13	0.13	0.09	0.09	0.05	0.07	0.05
Postdoc total appointments (in any sector)	0.00	0.02	0.01	0.02	0.02	0.04	0.03
Other academic position	0.01	0.01	0.06	0.06	0.06	0.07	0.09
Industry	0.13	0.12	0.12	0.18	0.26	0.23	0.25
Federal labs and other government	0.11	0.11	0.12	0.09	0.09	0.09	0.08
Other jobs (including self-employed)	0.06	0.06	0.08	0.11	0.08	0.08	0.07
Research Involvement (in full-time S&E position)							
Engaged primarily in research	0.39	0.43	0.44	0.45	0.56	0.60	0.58
Supported by federal grants/contracts	0.47	0.46	0.40	0.39	n/a	0.40	0.40
Supported by HHS, NSF, and/or USDA	0.41	0.38	0.36	0.33	n/a	0.37	0.38

Table F.7N Career Progression of Biomedical Life-Science PhDs–US Citizens and Permanent Residents

	\multicolumn{7}{c}{Survey Year}
	1973	1977	1981	1985	1989	1993	1995
Panel A: Number in 1-2 Year Cohort							
Not in Full-time S&E Work Force							
Unemployed and seeking position	121	144	110	115	80	52	199
Part-time employed	157	117	174	131	154	182	124
Working outside science and engineering	91	164	52	183	129	215	168
Full-time employed in S&E field							
Tenure-track faculty position @ PhD Institutions	1321	926	774	394	576	619	546
Tenure-track faculty position @ Other Inst	616	465	287	347	180	249	291
Postdoc total appointments (in any sector)	1521	2557	3167	3060	3681	3309	4348
Other academic position	487	215	461	501	517	740	900
Industry	401	419	562	642	521	797	524
Federal labs and other government	441	309	208	257	334	295	256
Other jobs (including self-employed)	363	264	303	389	292	228	182
Research Involvement (in full-time S&E position)							
Engaged primarily in research	3073	3335	4283	3894	4769	4588	5084
Supported by federal grants/contracts	2974	2968	3405	2675	n/a	2902	3620
Supported by HHS, NSF, and/or USDA	2573	2501	3095	2218	n/a	2637	3379
Panel B: Number in 3-4 Year Cohort							
Not in Full-time S&E Work Force							
Unemployed and seeking position	66	136	102	98	62	47	141
Part-time employed	135	90	102	103	211	265	229
Working outside science and engineering	157	144	82	61	118	318	370
Full-time employed in S&E field							
Tenure-track faculty position @ PhD Institutions	1835	1727	1793	1199	1059	966	892
Tenure-track faculty position @ Other Inst	671	655	367	420	155	434	427
Postdoc total appointments (in any sector)	349	815	1370	1538	1671	2061	2102
Other academic position	320	310	463	568	803	686	839
Industry	408	565	756	1311	1159	1064	1127
Federal labs and other government	421	399	395	425	383	298	385
Other jobs (including self-employed)	373	570	327	483	375	274	216
Research Involvement (in full-time S&E position)							
Engaged primarily in research	2039	2486	3707	3694	4156	4052	3998
Supported by federal grants/contracts	2400	2524	2647	2392	n/a	2664	2708
Supported by HHS, NSF, and/or USDA	2009	2055	2356	1999	n/a	2420	2502

	Survey Year						
	1973	1977	1981	1985	1989	1993	1995

Panel C: Number in 5-6 Year Cohort

Not in Full-time S&E Work Force

Unemployed and seeking position	31	111	117	54	67	87	71
Part-time employed	121	181	124	267	94	184	274
Working outside science and engineering	72	202	159	184	220	224	390

Full-time employed in S&E field

Tenure-track faculty position @ PhD Institutions	1692	2400	2064	1953	1482	1599	1369
Tenure-track faculty position @ Other Inst	464	717	515	338	471	334	589
Postdoc total appointments (in any sector)	78	324	514	713	782	893	958
Other academic position	163	281	407	458	720	527	799
Industry	395	631	873	1318	1538	1352	1224
Federal labs and other government	389	551	469	517	397	562	354
Other jobs (including self-employed)	272	532	493	431	470	395	358

Research Involvement (in full-time S&E position)

Engaged primarily in research	1523	2538	2842	3257	3735	3819	3619
Supported by federal grants/contracts	1732	2632	2550	2108	n/a	2166	2357
Supported by HHS, NSF, and/or USDA	1468	2164	2226	1862	n/a	2025	2158

Panel D: Number in 7-8 Year Cohort

Not in Full-time S&E Work Force

Unemployed and seeking position	36	71	51	63	28	36	88
Part-time employed	70	95	164	133	176	257	192
Working outside science and engineering	66	162	134	244	308	285	204

Full-time employed in S&E field

Tenure-track faculty position @ PhD Institutions	1267	2074	2282	2289	2081	1944	2032
Tenure-track faculty position @ Other Inst	391	666	530	414	371	277	362
Postdoc total appointments (in any sector)	40	162	183	119	393	406	407
Other academic position	69	136	278	425	626	646	498
Industry	313	507	829	1076	1530	1368	1347
Federal labs and other government	281	501	465	473	435	449	551
Other jobs (including self-employed)	169	411	579	477	594	405	441

Research Involvement (in full-time S&E position) 6073

Engaged primarily in research	1026	1894	2392	2823	3838	3399	3578
Supported by federal grants/contracts	1279	2088	2146	2153	n/a	2337	2249
Supported by HHS, NSF, and/or USDA	1087	1745	1740	1964	n/a	2191	2047

Appenidx F

	\multicolumn{7}{c}{Survey Year}
	1973	1977	1981	1985	1989	1993	1995
Panel E: Number in 9-10 Year Cohort							
Not in Full-time S&E Work Force							
Unemployed and seeking position	17	56	104	69	41	68	44
Part-time employed	61	117	149	168	231	147	201
Working outside science and engineering	72	168	162	206	238	400	241
Full-time employed in S&E field							
Tenure-track faculty position @ PhD Institutions	1085	1811	2382	2082	2272	1917	2178
Tenure-track faculty position @ Other Inst	296	511	498	503	311	453	305
Postdoc total appointments (in any sector)	4	59	53	122	97	230	175
Other academic position	29	40	296	339	401	461	547
Industry	302	480	631	981	1610	1433	1533
Federal labs and other government	249	443	610	481	565	567	489
Other jobs (including self-employed)	144	243	397	602	527	481	450
Research Involvement (in full-time S&E position)							
Engaged primarily in research	824	1555	2152	2311	3227	3321	3321
Supported by federal grants/contracts	981	1644	1956	2012	n/a	2237	2297
Supported by HHS, NSF, and/or USDA	873	1369	1734	1688	n/a	2066	2157

Table F.8 Number of Citizen and Permanent Resident Life-Science PhDs by sector, 1973-1995

	1973	1977	1981	1985	1989	1993	1995
Unemployed and seeking	343	869	879	1063	852	1414	1682
Part-time employment	809	940	1463	2164	3016	4581	4447
Working outside S&E	951	1775	1606	3392	4355	6378	6457
Tenure-track faculty: PhD-granting inst.	13685	18957	23923	27838	31857	32908	34257
Tenure-track faculty: other inst.	4817	6316	7469	8048	8191	9272	10014
Postdoctoral appointments	2202	4402	5772	6461	7567	8316	9851
Other academic appointments	1401	1438	3062	4251	5734	6532	7828
Industry	3547	5938	8845	13308	17724	20517	21185
Federal labs & other gov't	4406	6174	7503	8474	10160	10675	11143
Self-employed & others	1835	3117	4550	6035	7196	7733	8120
Total PhDs in workforce	33996	49926	65072	81034	96652	103920	114984

Appendix G

GETTING STARTED ON THE WORLD WIDE WEB: WEB SITES OF INTEREST TO YOUNG SCIENTISTS

This list is a starting point for readers who wish to search the Internet for information relevant to this report. It is neither complete nor fully representative. New sites open daily, and sites are discontinued without notice. Inclusion in the following list does not necessarily imply endorsement by the committee of the information found at the site.

Sites with a Focus on Young Scientists

- National Academy of Sciences Career Planning Center:
    http://www.nas.edu/cpc/index.html

- Science's Next Wave:
    http://nextwave.org

- Young Scientists' Network:
    http://www.edoc.com/jrl-bin/wilma/ema.800726377.htm [electronic newsletter]
    http://www.physics.uiuc.edu/ysn/ [archives and other information]

- Network of Emerging Scientists:
    http://pegasus.uthct.edu/nes/nes.html

- Pandora Science Policy Site:
    http://www.mit.edu:8001/afs/athena.mit.edu/user/e/r/erw/public/pandoralid.html

Networking and Education Sites

- Networking on the Network:
    http://weber.ucsd.edu/~pagre/network.html

- Principles of Protein Structure:
    http://www.cryst.bbk.ac.uk/pps/index.html
    http://pdb.pdb.bnl.gov/pps/index.html (us mirror site)

- Globewide Network Academy:
    http://www.gnacademy.org

- The Glyocoprotein Network and electronic conferencing:
    http://bellatrix.pcl.ox.ac.uk/tgn/Welcome.html

- BioMOO:
    http://bioinfo.weizmann.ac.il/biomoo

Scientific Societies

- National Academy of Sciences (NAS):
    http://www.nas.edu

- Federation of American Societies for Experimental Biology (FASEB):
    http://www.faseb.org

- American Association for the Advancement of Science (AAAS):
    http://www.aaas.org

- Association for Women in Science (AWIS):
    http://www.awis.org

- American Psychological Association (APA):
    http://www.apa.org

- American Society for Microbiology (ASM):
    http://www.asmusa.org

- Society for Neuroscience (SN):
    http://www.sfn.org

- American Society for Cell Biology (ASCB):
    http://www.ascb.org

- Links: The World of Science (hot links to American Chemical Society, and so on)
    http://www.annurev.org/general/univrsty.htm

- The World Wide Web Virtual Library of Biology Societies and Organizations:
    http://golgi.harvard.edu/afagen/depts/orgs.html

- American Society of Plant Physiologists
    http://aspp.org

National and Local Organizations of Graduate Students and Postdoctoral Fellows

- Postdoctoral Scientists Association (PSA) at University of California, San Francisco:
  http://saa49.ucsf.edu/psa/

- National Association of Graduate-Professional Students:
  http://nagps.varesearch.com/nagps/nagps-hp.html

Government Sites

- National Institutes of Health:
  http://www.nih.gov/

- National Science Foundation :
  http://www.nsf.gov/

- U.S. Dept.of Agriculture:
  http://www.usda.gov/

- Department of Energy:
  http://www.doe.gov/

Career Information (job listings and related information)

- Alternative Careers in Biosciences:
  http://www.mbb.yale.edu/acb/

- Bioweb Career Center:
  http://www.bioweb.com/

- Bio OnLine:
  http://www.bio.com/hr/hr_index.html

- Employment Links for the Biomedical Student:
  http://www.medcor.mcgill.ca/expmed/docs/elbs.html

- Education and Careers in Science and Technology–Alfred Sloan Foundation:
  http://www.sloan.org/education/index.html

- MedSearch America (mostly health care-related fields):
  http://www.medsearch.com/

- Job Listings and Career Services (Biosciences):
    http://golgi.harvard.edu/biopages/jobs.html

- 200 Links to Web sites that describe specific careers after training in biology:
    http://www.furman.edu/~snyder/careers/careerlist.html

- Nature:
    http://www.nature.com

- Survival Skills and Ethics; University of Pittsburgh:
    http://www.pitt.edu/~survival/homepg.html

News groups. forums for discussion

- news:sci.research.postdoc

- news:sci.research.careers